Visualizing Differential Equations with Slope Fields

by Alan Lipp

Publisher: Tom Maksym

Executive Editor: Steven Jay Griffel

Vice President of Production and Manufacturing: Doreen Smith

Production Manager: Jason Grasso

Project Manager: Matthew Hjembo

Senior Book Coordinator: Paul Zakrzewski

Production Assistant: Richard Lehmbeck

Designer: Eric Dawson

Copy Editor: Dee Josephson

Proofreader: Pat Smith

Technical Art: Matthew Hjembo, Richard Lehmbeck, Alan Lipp, Sharon MacGregor, Paul Zakrzewski

PEOPLES PUBLISHING GROUP ISBN 1-4138-1322-4

The Peoples Publishing Group, Inc.
299 Market Street
Saddle Brook, New Jersey 07663

Printed in the United States of America.

10 9 8 7 6 5 4 3 2 1

Library of Congress Cataloging-in-Publication Data

Lipp, Alan.
Calculus visualizing differential equations with slope fields / by Alan Lipp.
p. cm. -- (Little books of big ideas)
ISBN 1-4138-1322-4
1. Calculus--Graphic methods. 2. Vector fields. 3. Calculus--Problems, exercises, etc. 4. TI-83 (Calculator) 5. TI-84 (Calculator) I. Title. II. Little books of big ideas (Peoples Publishing Group)

QA303.2.L57 2005
515--dc22

2005000310

Dedication

To my father who showed me the road, to Edie who encouraged me to walk it,
and to my students who have always been my best teachers.

TABLE OF CONTENTS

Acknowledgments

The writing of a book, any book, is always a group effort as the writing impacts and is impacted by many people in the author's life. This book is no exception. So first I must acknowledge my endless gratitude to my wife who was willing to have me disappear behind a computer screen for months at a time.

Boundless thanks to Dr. James Henle who helped me produce the graphics in the text using software routines that he created and was selflessly willing to share with me. Thanks too to Anne Murphy who offered invaluable feedback on content and to Lynn Quitman Troyka who provided support and encouragement and bountiful advice on writing and publishing.

Introduction

Technology has transformed the teaching of calculus. The availability of high-powered computers and graphing calculators has allowed new topics to enter the curriculum and made new tools available to visualize calculus.

Slope fields were introduced into the BC Calculus curriculum in 1998, and questions on slope fields first appeared on the AB Calculus exam in May 2004. As a recent entry to the calculus curriculum, slope fields are not well covered in some textbooks and are unfamiliar to many calculus teachers. This book is intended to help fill that gap. It is suitable for calculus teachers in high school and college, and first-year calculus students of any age.

The pace of the text is gentle and the style is informal. It is meant to be readable. Every effort has been made to introduce ideas slowly and carefully, with many worked examples and many problems for practice. The text also offers many hints and cautions in order to avoid common misconceptions that can arise through improperly reading the graph of a slope field.

Students learn best when they can connect new information to prior knowledge. So you will find suggestions throughout the text connecting slope fields to topics studied earlier in the year such as sketching a function, given the graph of its derivative, or implicit differentiation.

Each chapter is short and limited in scope. The intent is to make each one accessible as a single lesson. The chapters include a problem set with problems practicing the skills presented, as well as more thought-provoking questions extending the ideas into new regions. Because many problems require students to draw on a slope field, all graphics included in this volume can be downloaded at www.peoplescollegeprep.com/slopefields.html so that students and teachers will have consumable copies of these figures for use in the classroom and for homework.

Chapter 1 presents the first examples of slope fields and probes their use before they are more formally defined in Chapter 2, where there are explicit instructions on constructing slope fields by hand. Readers will find many suggestions for thinking about these graphics and how to extract information from them. A program for constructing slope fields with the TI-83 and TI-84 series of calculators is provided in an Appendix as well.

Chapter 3 uses slope fields as an aid in constructing solutions to initial value problems. Chapter 4 focuses on common errors that arise when attempting to find a solution to a problem where the derivative fails to exist at one or more points. Chapter 5 introduces the notion of equilibria and stable and unstable equilibrium solutions. It examines how slope fields can be used to help visualize these solutions.

Chapter 6 focuses on derivatives that are functions of a single variable, either x or y. This is a natural extension of ideas presented early in differential calculus but usually not revisited until the study of logistic equations. However, this technique has broader applicability, and that is explored here. The solution of logistic equations is covered informally and can serve as an introduction to that topic.

Chapter 7 introduces Euler's Method of approximating solutions to initial value problems. The chapter reinforces the idea of tangent-line approximations. Both a calculator program and instructions for using a spreadsheet to implement Euler's Method are provided in the Appendices. Chapter 8 offers a final review, a collection of problems covering the earlier chapters, followed by a self-assessment test.

Calculus can be an exciting and heady intellectual adventure when you have good tools available to help. It can also be frustrating and daunting when access to the ideas is masked by dense language and confusing graphics. This book will be a success if you find the writing clear and the visualizations helpful. I would love to hear from you in either case. I welcome your comments and criticisms, as they will help to improve future editions. You can reach me at alipp@williston.com.

Alan Lipp

South Deerfield, Massachusetts

CHAPTER 1

What Are Slope Fields?

In this book you will explore special types of graphs that can be helpful in visualizing the solutions to differential equations. Remember that, although a differentiable function has only one derivative, there are infinitely many functions that share the same derivative, because shifting a function vertically does not change its derivative. Each such function is called a **particular solution** of the differential equation. The **general solution**, however, includes one or more arbitrary constants and represents the infinite family of functions that solve the differential equation. A **slope field** is a graphical way of picturing particular solutions as well as the general solution.

Example 1 Describe the family of solutions to the differential equation $\frac{dy}{dx} = 2x^2$.

Solution Antidifferentiate both sides to get $y = \frac{2}{3}x^3 + C$. This is the general solution, a family of cubic functions that are all vertical shifts of the particular solution $y = \frac{2}{3}x^3$.

All is straightforward when the derivative is a sufficiently simple function of x so that we can find the antiderivative directly, as in the foregoing example. Often, however, we cannot do so directly, such as when the derivative is a more complex function of x. In these cases we can still determine a general solution using the **Second Fundamental Theorem of Calculus.**

Example 2 Describe the family of solutions to the differential equation $\frac{dy}{dx} = \cos(x^2)$.

Solution There is no simple formula for the antiderivative of $\cos(x^2)$. However, we can use the Second Fundamental Theorem of Calculus to write an antiderivative. The integral expression $y(x) = \int_a^x \cos(t^2)dt$ defines a function of x and represents the particular solution passing through $(a, 0)$. The general solution is the family of functions generated by varying the constant a. Alternatively, we could choose a particular value of a, say $a = 0$, and then the general solution could be written as $y(x) = \int_0^x \cos(t^2)\,dt + C$.

Although the Second Fundamental Theorem gives us a way to write an expression for the general solution, it is not very satisfying, because these integrals do not look like nice algebraic formulas. That cannot be helped; sometimes there is no nice algebraic formula. However, it is still possible to get a feel for these integral solutions. This is where the central topic of this book comes in. Slope fields are a way to visually represent solutions to differential equations. Much more on slope fields will be explored in the rest of the book, including how to make your own slope fields. For now it will be enough to observe several slope fields.

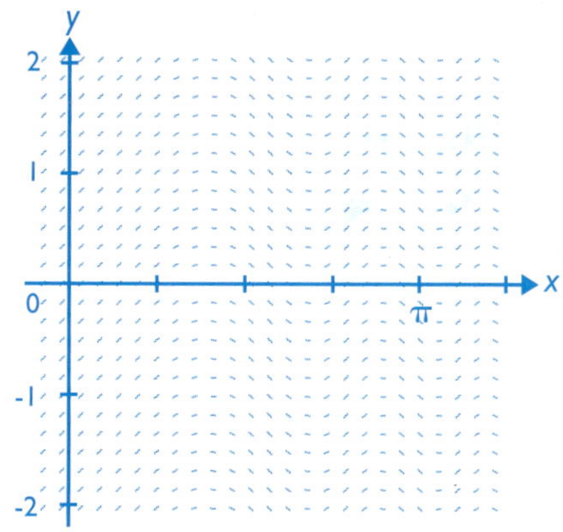

Figure 1.1

Figure 1.1 shows a slope field for the differential equation $\frac{dy}{dx} = \cos(x^2)$. The slope field is composed of hundreds of tiny line segments. Taken together, they suggest many waves moving off to the right. If the slope field were a stream, a light object dropped at any point (x_0, y_0) would be carried away "downstream" along these waves.

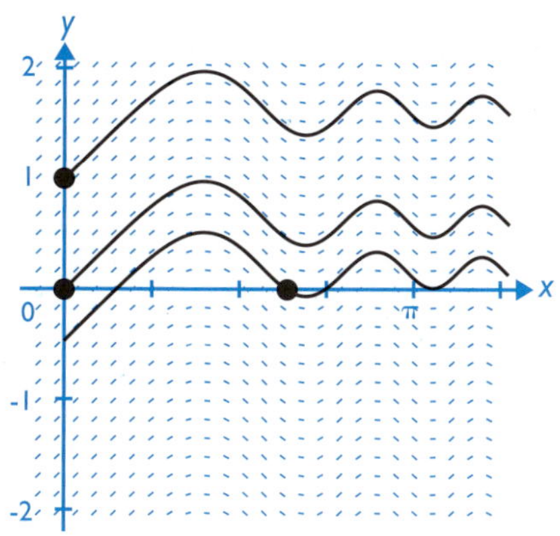

Figure 1.2

The slope field in Figure 1.1 suggests something about the general solution to the differential equation, whereas the three curves in Figure 1.2 approximate particular solutions through (0, 1), (0, 0), and (2, 0).

Notice that these waves appear to be vertical shifts of each other. Notice, too, how the slope field suggests that successive periods of the wave appear to shorten and diminish in amplitude as we move to the right. Although we still do not have a nice formula for these curves, this slope field gives a much better sense of some of the properties of the solutions to $\frac{dy}{dx} = \cos(x^2)$.

The Second Fundamental Theorem of Calculus, together with the use of slope fields, allow us to solve differential equations algebraically (by formula) and to approximate graphical solutions as well. There are times, however, when even the Second Fundamental Theorem cannot help. If the derivative is an **implicit function** (a function involving y), then we cannot compute values of the solution using the Second Fundamental Theorem. However, we can still visualize the family of solutions by using a slope field.

Example 3 Contrast the particular solutions to the differential equation

$$\frac{dy}{dx} = \frac{2x - 2xy}{3y^2 + x^2 - 2y - 16} \text{ passing through } (0, 6), (0, 0), \text{ and } (0, -5).$$

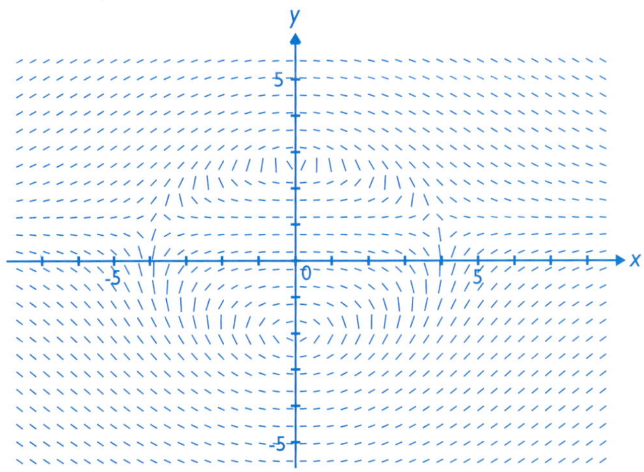

Figure 1.3a

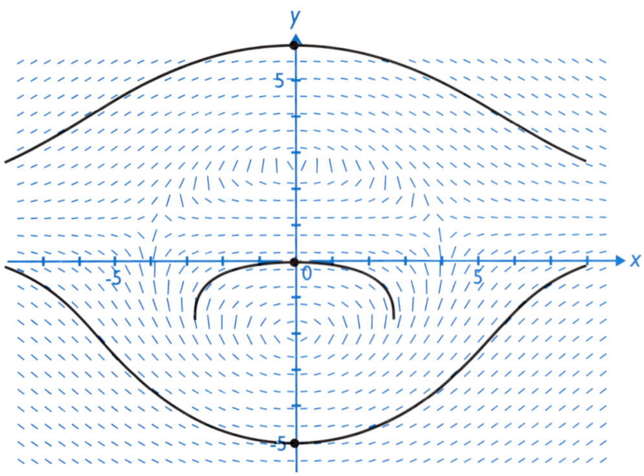

Figure 1.3b

Solution

We cannot antidifferentiate this equation directly because the right-hand side is a function of both x and y. We cannot use **Separation of Variables** to solve this problem algebraically, nor can we use the Fundamental Theorem of Calculus to write an integral expression for the solution. There are other methods for approximating solutions, and we will examine one in a later chapter, but for now a slope field is the only tool we have to try to learn something about the solutions. The slope field shown in Figure 1.3a may seem quite messy. It will take some practice before you will be able to get much information from a picture this complex, but even here we can learn quite a bit about the solutions to this differential equation. In Figure 1.3b we have drawn solutions passing through the three given points. Notice that these three solution curves are not simply vertical shifts of each other, as in Example 2. Instead, the slope field shows that this equation has several distinct types of solutions with different families of solutions in different parts of the plane.

The curves passing through (0, 6) and (0, −5) appear to have the domain of all real numbers. The first has a **relative maximum** on the y-axis while the second has a **relative minimum** on the y-axis. Each curve appears to have two **points of inflection**. The solution passing through (0, 0), however, seems to have a limited domain and to be **concave down** throughout.

Look again at the slope field in Figure 1.3a. Do you now see how the top portion of the field suggests curves similar to the solution passing through (0, 6)? Likewise, the bottom of the field suggests a family of curves similar to the solution passing through (0, −5). But if you look near the origin there appear to be two sets of loops, like eddies in a stream. The curve passing through (0, 0) is the top half of one of these eddies. Why do you suppose we draw only the top half of the loop and not the entire loop as our particular solution? We will ask this question again in the problems at the end of the chapter.

Interestingly, this slope field suggests other properties of the solutions to the differential equation as well. For example, *all* the solutions appear to be symmetrical across the y-axis.

Problems

www.peoplescollegeprep.com/slopefields.html

1. Figure 1.4 gives a slope field for the differential equation $\frac{dy}{dx} = xe^{-x^2}$.

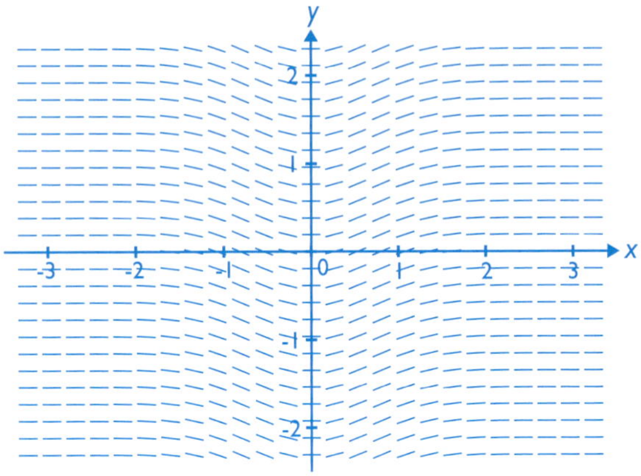

Figure 1.4

a. How many types of solutions do there appear to be?

b. Sketch the solution passing through $(0, -2)$. Where are the locations of the turning points of all the solutions? Do the solutions appear to have points of inflection? If so, how many?

c. Find the general solution by antidifferentiation and graph the particular solution passing through $(0, -2)$. Does your sketch confirm your answers in **1a** and **1b**?

2. In Example 2 it was stated that there is no simple antiderivative of $\cos(x^2)$. Show that $\sin(x^2)$ is not a correct antiderivative.

3. Figure 1.5 gives a slope field for the differential equation $\frac{dy}{dx} = x\sin(x^2)$.

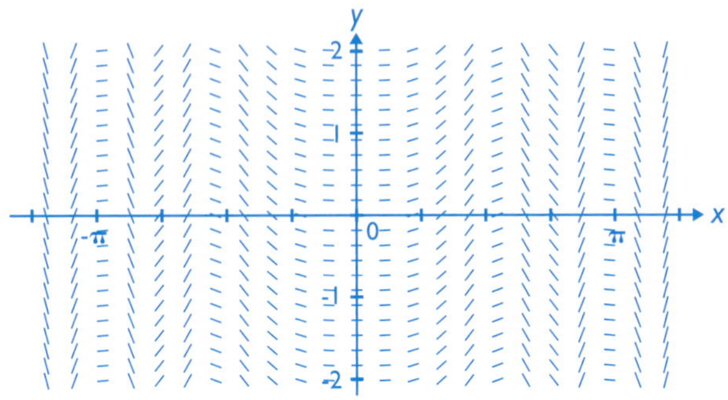

Figure 1.5

a. How many types of solutions do there appear to be? Do the solutions appear to be **even, odd,** or neither?

b. Find the general solution by antidifferentiation and then graph the particular solution passing through $(0, 0)$. Does your graph confirm your answers in **3a**?

4. Figure 1.6 gives a slope field for the differential equation $\frac{dy}{dx} = 2x - 3$.

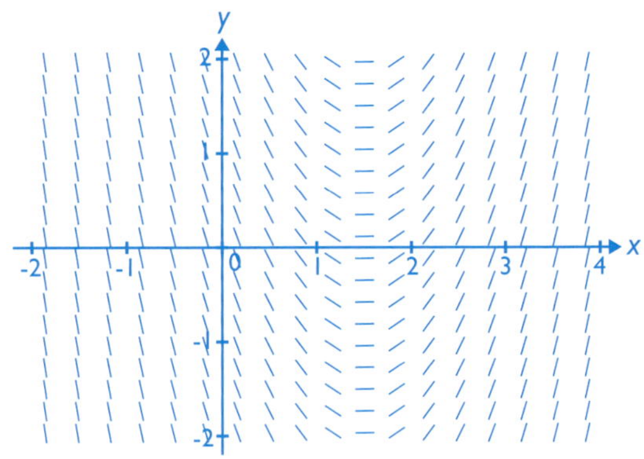

Figure 1.6

a. Do the solutions appear to be even, odd, or neither? Do the solutions appear to be concave up or concave down? What kind of curves do the solutions seem to be?

b. Find the general solution by antidifferentiation. Does your answer confirm your answers in **4a**?

5. Figure 1.7 gives a slope field for the differential equation $\frac{dy}{dx} = \frac{1.5}{1 + x^2}$.

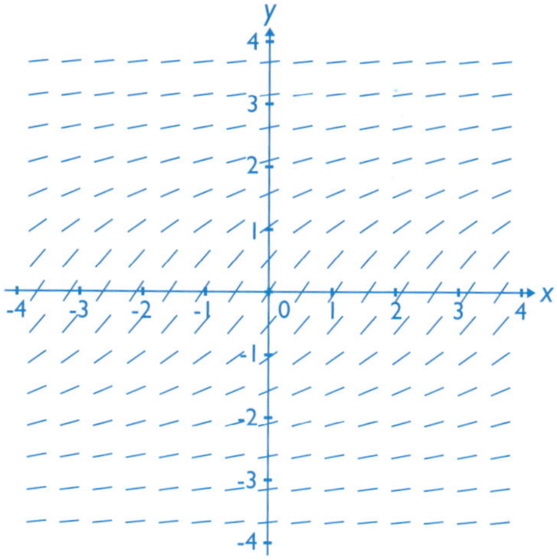

Figure 1.7

a. Do the solutions appear to be even, odd, or neither? Do they appear to have points of inflection and, if so, where? On what domain do the solutions appear to be increasing?

b. Find the general solution by antidifferentiation and then graph the particular solution passing through (1, 1). Does your graph confirm your answers in **5a**?

6. Figure 1.8 gives a slope field for the differential equation $x^2\frac{dy}{dx} = x - \sqrt{x}$. The solution passing through (2, 1) is shown.

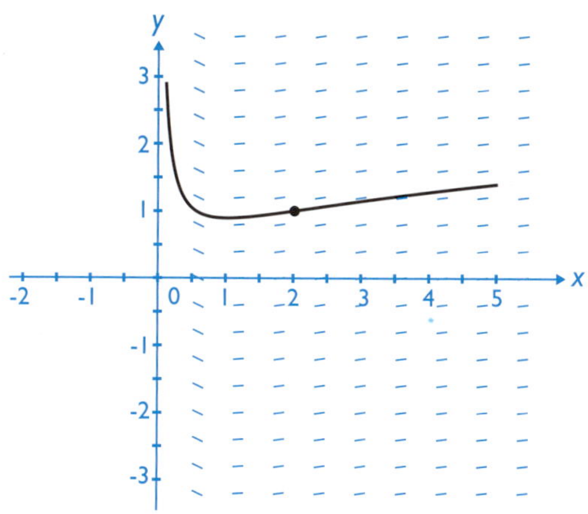

Figure 1.8

a. Sketch the solution passing through (3, −1.5) and use it to approximate $y(4)$.

b. What appears to be the domain of the solutions? Explain why that is reasonable.

c. Do the solutions appear to have points of inflection and, if so, where? On what domain do the solutions appear to be increasing?

d. Find the general solution by antidifferentiation, and then graph the particular solution passing through (3, −1.5). Does your graph confirm your answers above?

7. Figure 1.9 gives a slope field for the differential equation $x^2\frac{dy}{dx} = x\frac{dy}{dx} - x^2$.

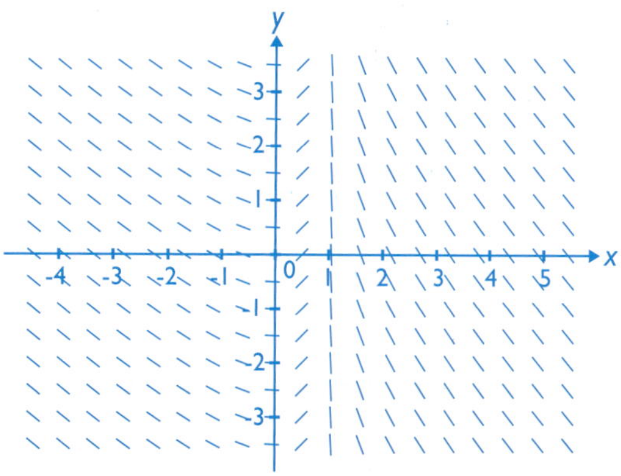

Figure 1.9

a. What appears to be the domain of the solutions?

b. What seems to be happening near $x = 1$ in the slope field? Can you explain how the differential equation could be used to support your answer?

c. Find the general solution and then graph the particular solutions passing through (2, 0) and through (−2, 0). Do your graphs confirm your answers in **7a**?

8. In Example 3 we drew only the top half of the loop passing through (0, 0). Explain why it would be incorrect to draw the entire closed loop as a solution.

9. Example 3 showed three types of solution curves to $\frac{dy}{dx} = \frac{2x - 2xy}{3y^2 + x^2 - 2y - 16}$. There are actually six different families of solutions.

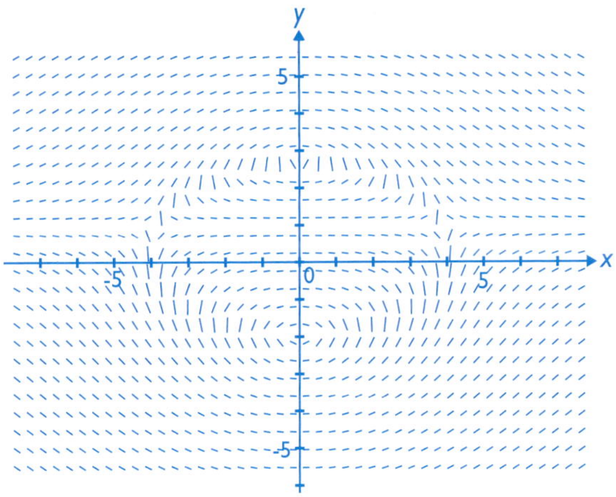

Figure 1.10

On a copy of the slope field, shown here in Figure 1.10, find three new types of solution curves. For each new type you find:

a. Give a possible initial value and sketch a particular solution.

b. Describe the important features of the curve you drew.

10. Figure 1.11 gives a slope field for the differential equation $2x + 2y\frac{dy}{dx} = 0$.

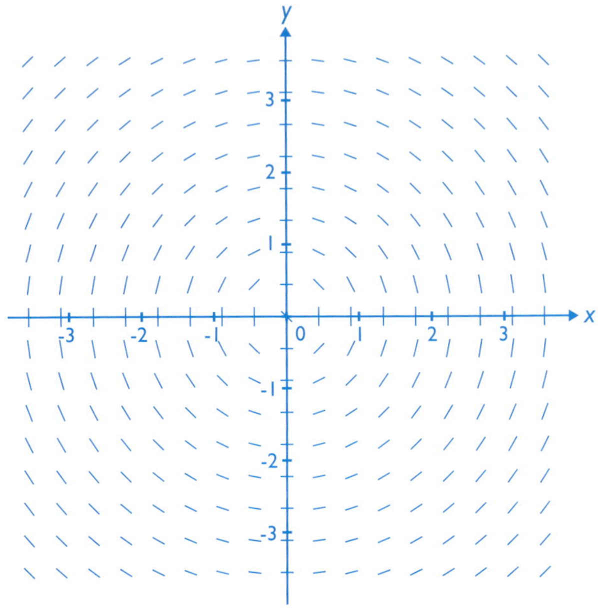

Figure 1.11

a. How many types of solutions does this equation appear to have?

b. Guess an implicit function, which, when differentiated, gives the equation.

c. Use your answer to **10b** to find the equation of the solution through (3, −4).

11. Figure 1.12 gives a slope field for the implicit differential equation $y + x\frac{dy}{dx} = 0$.

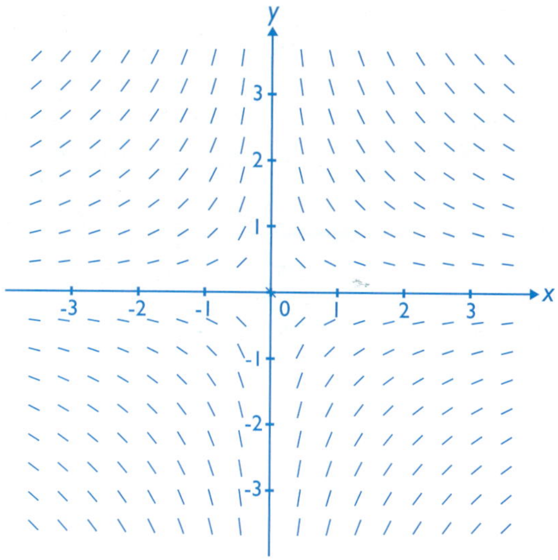

Figure 1.12

a. How many types of solutions does this equation appear to have?

b. Guess an implicit equation, which, when differentiated, gives this derivative. (*Hint*: the left-hand side is the derivative of a product.)

12. Figure 1.13 gives a slope field for the implicit differential equation

$$x - 2e^x - 2y\frac{dy}{dx} = x\frac{dy}{dx} + y.$$

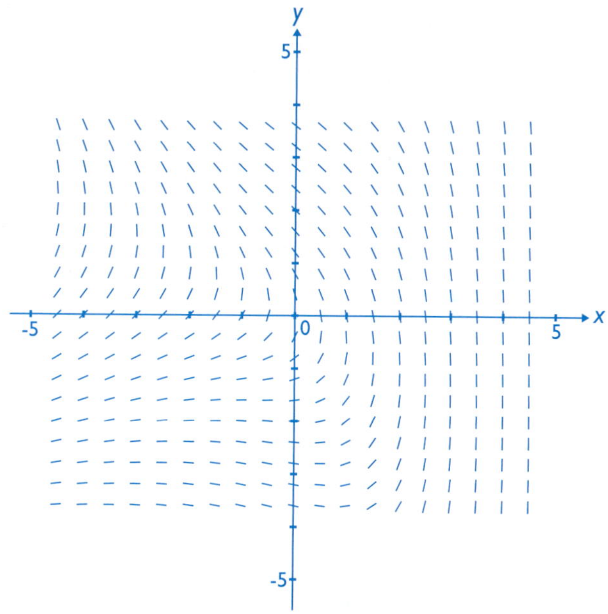

Figure 1.13

a. How many types of solutions does this equation appear to have?

b. Find the general solution to this implicit differential equation. (*Hint:* Does the right-hand side seem similar to the equation in Exercise 11?)

CHAPTER 2

Constructing Slope Fields

As we saw in the first chapter, slope fields are visualizations of solutions to differential equations. The slope field of a simple equation like $\frac{dy}{dx} = x^2$ consists of a family of vertically shifted cubic curves $y = \frac{x^3}{3} + C$. The slope fields of implicit differential equations can be much more complicated, having many different types of solutions. But as we saw, even these slope fields can tell us quite a bit about solutions to an equation that we may not be able to solve algebraically. In this chapter you will learn how to draw slope fields and how to recognize some of their important features.

To begin, consider the simple differential equation $\frac{dy}{dx} = \frac{x}{2}$. Instead of thinking of it as a problem to be solved, use it as a recipe for computing slopes. For any given value of x we can compute a value, and this value represents the slope of a solution curve at a particular point. Such computations are exactly how you first used derivatives to find tangent lines of curves.

Since the general solution of a differential equation consists of a family of curves, there will be solutions passing through every point (x, y) in the domain of the derivative. At each of these points we can use the differential equation to compute a value of the derivative. Table 2.1 gives the slopes $\frac{dy}{dx} = \frac{x}{2}$ at twenty-five different points.

x \ y	-2	-1	0	1	2
-2	-1	-1	-1	-1	-1
-1	$-\frac{1}{2}$	$-\frac{1}{2}$	$-\frac{1}{2}$	$-\frac{1}{2}$	$-\frac{1}{2}$
0	0	0	0	0	0
1	$\frac{1}{2}$	$\frac{1}{2}$	$\frac{1}{2}$	$\frac{1}{2}$	$\frac{1}{2}$
2	1	1	1	1	1

Table 2.1

Note that these are not all slopes of the same particular solution curve. Some of these might be on the same curve or each of these twenty-five values could be on a different particular solution. The computed values are independent of the y-value, because the derivative is a function of x only.

We can use such a table of values to find a tangent line at each of the points (x, y) we choose. If we then draw a small segment for each of the many different tangent lines, the resulting picture is a slope field.

Example 1 Make a slope field for the differential equation $\frac{dy}{dx} = x$.

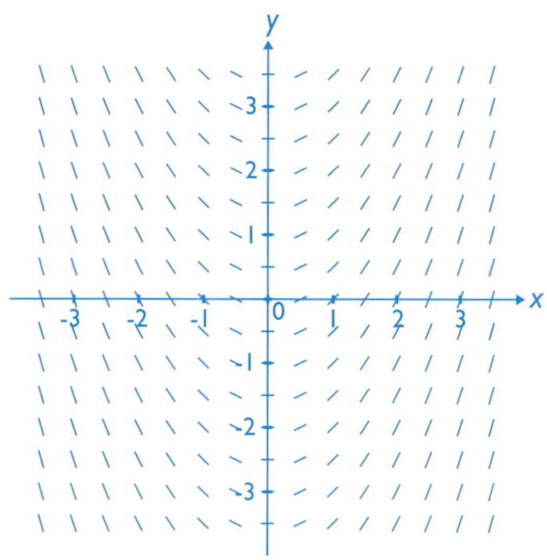

Figure 2.1

Solution Figure 2.1 shows a slope field for this differential equation. On it we have drawn many tiny line segments, each one a small portion of a tangent line. Notice how the segments in the diagram seem to flow around parabolas. These tangent segments help us visualize solutions to the differential equation.

Of course, in this example it would be easy to find the exact solution because we could antidifferentiate directly; but when we cannot find an antiderivative, slope fields can suggest features of the solution curves that cannot be seen in other ways.

Now, instead of looking at the parabolas, focus on the line segments along the line $x = 2$. These appear to be parallel. In fact, the segments along every vertical line appear to be parallel. Can you think why this should be so?

To draw a slope field, follow these three steps:

1. Choose a window of the plane in which to view the slope field. For Figure 2.1 we chose $[-3.5, 3.5] \times [-3.5, 3.5]$.
2. Choose the points at which to compute the slopes. This is usually done by choosing the spacing Δx and Δy between the points. In Figure 2.1, $\Delta x = \Delta y = 0.5$.
3. At each chosen point, compute the slope from the differential equation, and draw a small segment of the tangent line at each point.

To visualize the solution curves, we need hundreds of segments. There are 225 segments in Figure 2.1—far too many to draw by hand. In general we will use software to construct slope fields conveniently. (Slope field programs for several types of calculators are given in Appendix A.) Still, it is instructive to draw a few slope fields by hand to get a sense of the level of detail involved. The next example shows one way to do so.

Example 2 Draw a slope field for $\frac{dy}{dx} = -y$ in the window $-2 \le x \le 2$ and $-2 \le y \le 2$. Use $\Delta x = \Delta y = 1$ so that the tangents are drawn at points with integer coordinates.

x \ y	−2	−1	0	1	2
−2	2	1	0	−1	−2
−1	2	1	0	−1	−2
0	2	1	0	−1	−2
1	2	1	0	−1	−2
2	2	1	0	−1	−2

Table 2.2

Solution Begin by computing the slopes you will need, which are organized in Table 2.2. Because the slopes are independent of x, it is convenient to draw the slope field in horizontal rows, where all segments have the same slope. Begin, for example, with points along the line $y = -2$. For every value of x the derivative is $+2$, so at each of these five points we draw a short segment with slope 2, as in Figure 2.2a.

Continue by drawing segments with slope 1 along $y = -1$, segments with slope 0 along $y = 0$, segments with slope -1 along $y = 1$, and segments with slope -2 along $y = 2$. The completed slope field is shown in Figure 2.2b.

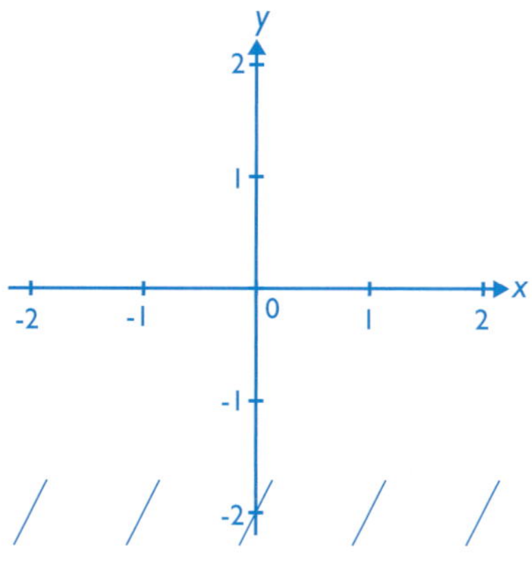

Figure 2.2a

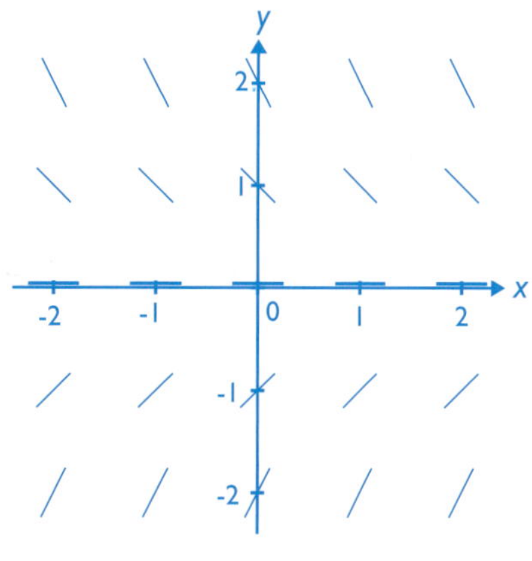

Figure 2.2b

Understanding how slope fields are drawn is the key to being able to read the important features of solution curves. In the next example, we use the properties of the derivatives to match slope fields to differential equations.

Example 3 Suppose the slope fields in Figures 2.3a–c were mislabeled with the wrong differential equations. Which is the slope field for $y' = xy$, which for $y' = x + y$, and which for $y' = 2x$?

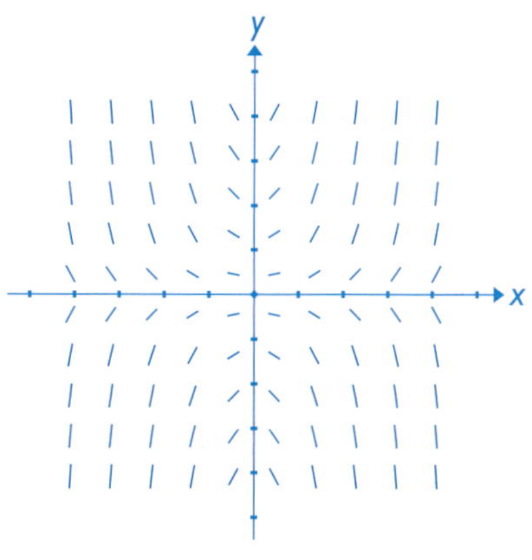

Figure 2.3a

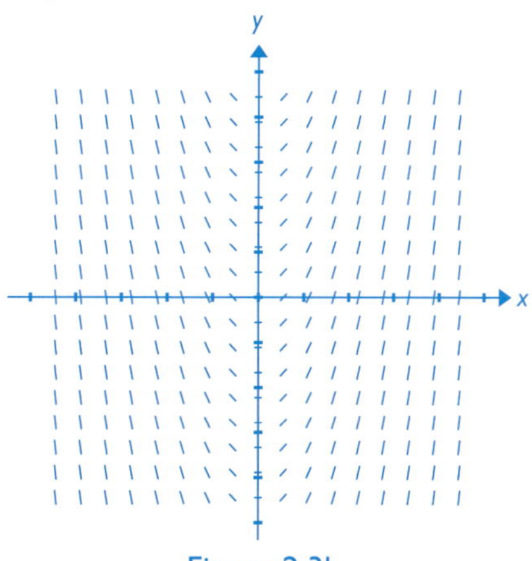

Figure 2.3b

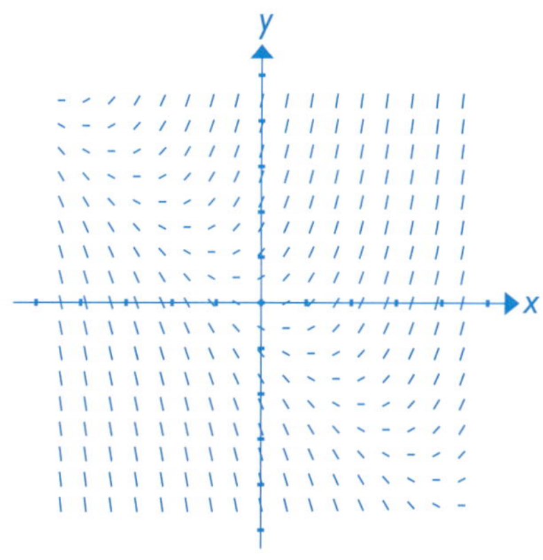

Figure 2.3c

Solution $y' = 2x$ does not depend upon y. Therefore, its solution curves will have the same slope along any vertical line. Scan each figure by looking up and down the columns of segments. Only Figure 2.3b shows the same slope throughout the column and so must be the slope field for $y' = 2x$.

The derivative $y' = xy$ will be positive in quadrants 1 and 3, where x and y have the same sign, and negative in quadrants 2 and 4, where their signs are different. The line segments in Figure 2.3c all have positive slope only in quadrant 1, and those in Figure 2.3b have positive slope in both quadrants 1 and 4. Because only Figure 2.3a has the correct signs in all four quadrants, it must be the slope field for $y' = xy$.

Of course, there is only one choice left, but it is instructive to ask why Figure 2.3c must show the slope field for $y' = x + y$. Notice that the derivative $x + y$ equals zero whenever $y = -x$. Look along the diagonal line $y = -x$ in all three slope fields. Because the tangent line segments are horizontal along $y = -x$ only in Figure 2.3c, Figure 2.3c is the only possible choice for $y' = x + y$.

Example 3 suggests several strategies to use when examining a slope field.

- Examine the variables involved in the derivative. If y' is a function of x only, then the slope field will have parallel tangent line segments along each vertical line. If y' is a function of y only, then the slope field will have parallel tangent line segments along each horizontal line.
- Examine the signs of the derivative, and compare those signs to the slopes of the tangent line segments in the slope fields.
- Examine the zeros of the derivative. The points where $y' = 0$ mark the location of horizontal tangents to solution curves. Look to see if this is represented in the slope field with horizontal tangent line segments.

Example 4 illustrates two other important features of some slope fields.

Example 4 Four different slope fields are given in Figures 2.4a–d. Determine which is the slope field for $y' = \frac{x}{(y-2)}$, which is for $y' = \sqrt{y}$, which is for $y' = \frac{2x}{(y-2)}$, and which is for $y' = \frac{2}{(x+2)}$.

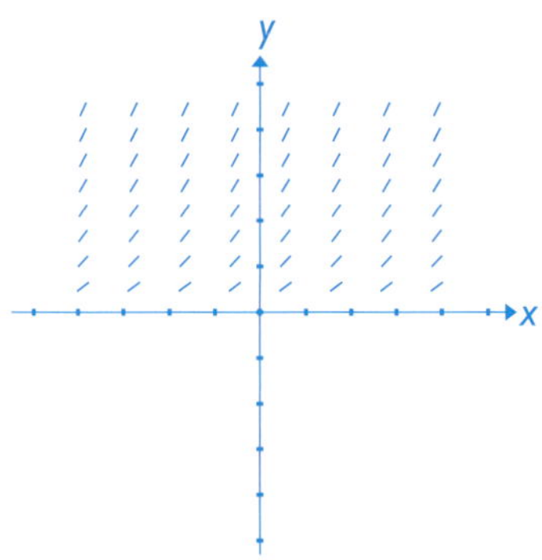

Figure 2.4a

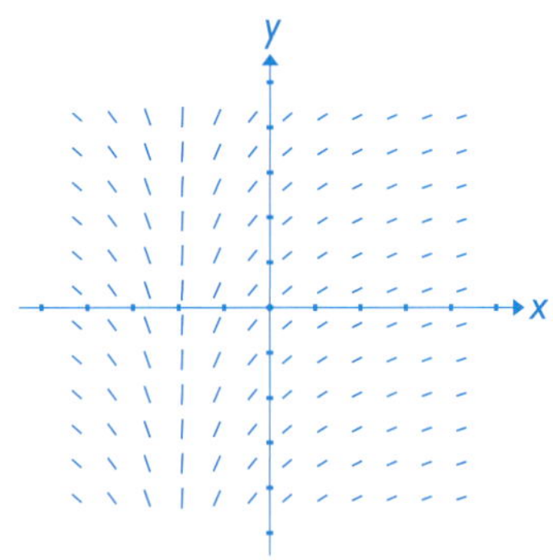

Figure 2.4b

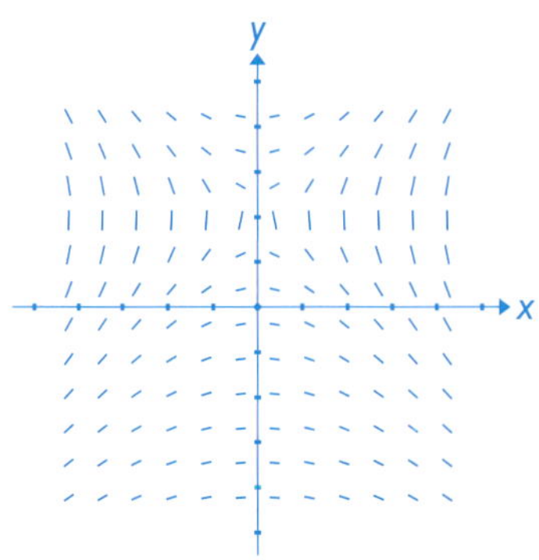

Figure 2.4c

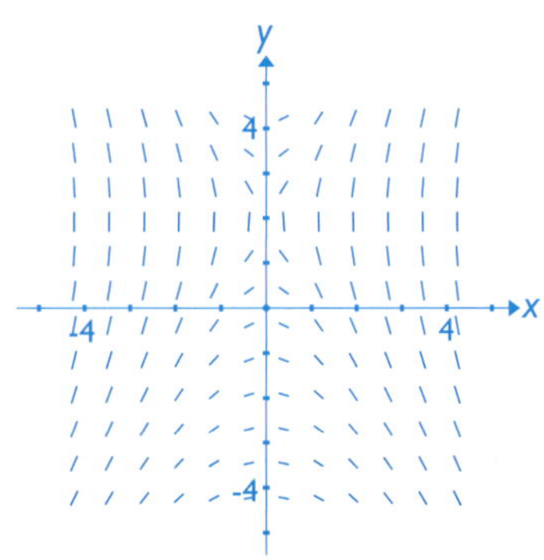

Figure 2.4d

Solution

Figure 2.4a shows a slope field with a restricted domain. Because there are no segments drawn below the x-axis, we know that the derivative exists only for $y > 0$ (or possibly for $y \geq 0$). Because $y' = \sqrt{y}$ is the only equation that places this restriction on y, it is matched with Figure 2.4a.

The equation $y' = \frac{2}{(x+2)}$ is undefined at $x = -2$. As x approaches -2 the values of y' increase rapidly, telling us to look for vertical or near-vertical tangent seg ments along $x = -2$. So Figure 2.4b must be the slope field for this differential equation. The remaining equations, $y' = \frac{x}{(y-2)}$ and $y' = \frac{2x}{(y-2)}$, are each unde fined along the horizontal line $y = 2$. As we near this line, the slopes become very steep. That is, as $y \to 2$, $y' \to \pm\infty$. Figures 2.4c and 2.4d each have hori zontal bands of vertical tangent lines along $y = 2$, but which belongs to which differential equation?

	$y' = \frac{x}{(y-2)}$	$y' = \frac{2x}{(y-2)}$
$(-3, 0)$	1.5	3
$(-1, -1)$	$\frac{1}{3}$	$\frac{2}{3}$

Table 2.3

The best strategy here is to pick a convenient point or two and compare the values of the derivatives to the tangent lines in the slope fields at these points. Several values of these equations are shown in Table 2.3.

Compare the tangent lines on each graph at these points. Because the lines in Figure 2.4d are steeper, Figure 2.4d is the slope field for $y' = \frac{2x}{(y-2)}$, and Figure 2.4c is the slopefield for the other derivative, $y' = \frac{x}{(y-2)}$.

Summary

Key features of slope fields to use in identifying them:

- The number of variables in the derivative;
- The signs of the derivative;
- The zeros of the derivative;
- The domain of the derivative;
- The location of the vertical tangents;
- The slopes at selected points.

In Appendix A of this book are slope field programs for several different calculators. Now might be a good time to enter one in your calculator.

Problems

www.peoplescollegeprep.com/slopefields.html

1. Sketch a slope field for $\frac{dy}{dx} = 1$. Draw your tangent segments on a copy of Figure 2.5 at coordinates (x, y) where x and y are each either -1, 0, or 1.

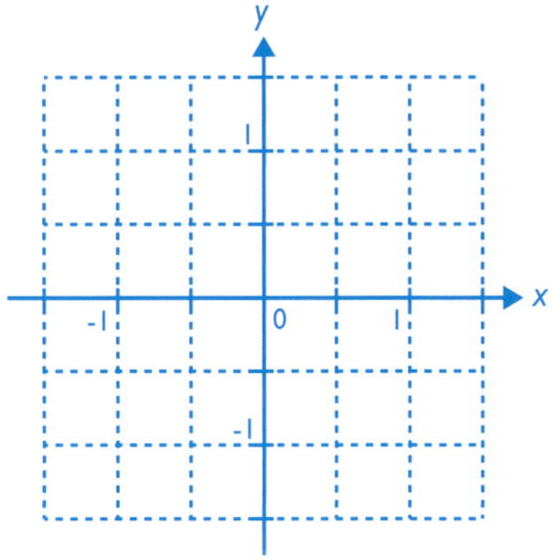

Figure 2.5

2. Sketch a slope field for $\frac{dy}{dx} = x - y$. Draw your segments on a copy of Figure 2.5 at coordinates (x, y) where x and y are each either -1, 0, or 1.

3. Figure 2.6a shows a graph drawn by absentminded Albert. He drew 10 tangent lines to a curve but forgot to draw the curve itself. On a copy of Figure 2.6a, draw in the curve that Albert forgot. Can you guess a formula for Albert's curve?

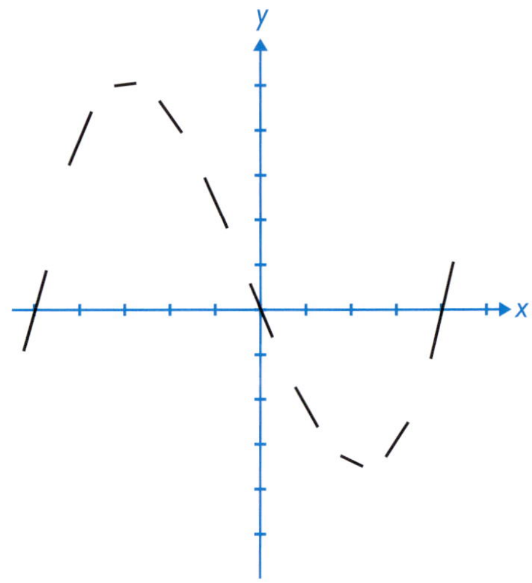

Figure 2.6a

4. Figure 2.6b shows another graph, this one drawn by absentminded Albert's sister Alice. She also drew 9 tangent lines but forgot to draw the curve itself. On a copy of Figure 2.6b draw in the curve that Alice left out. Can you guess a formula for Alice's curve?

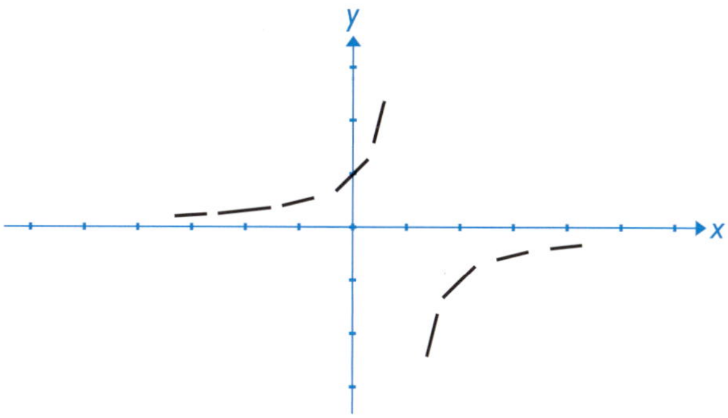

Figure 2.6b

5. Explain why Figure 2.7 could not be the slope field for $\frac{dy}{dx} = -yx^2$.

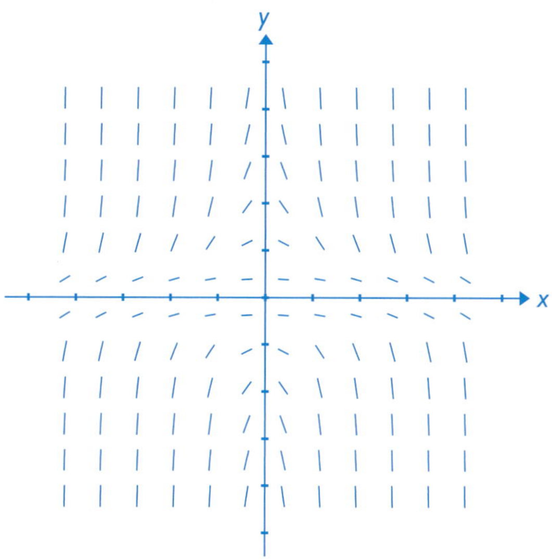

Figure 2.7

6. Figure 2.8 shows the slope field for the differential equation $\frac{dy}{dx} = \frac{x+y}{x^2}$.

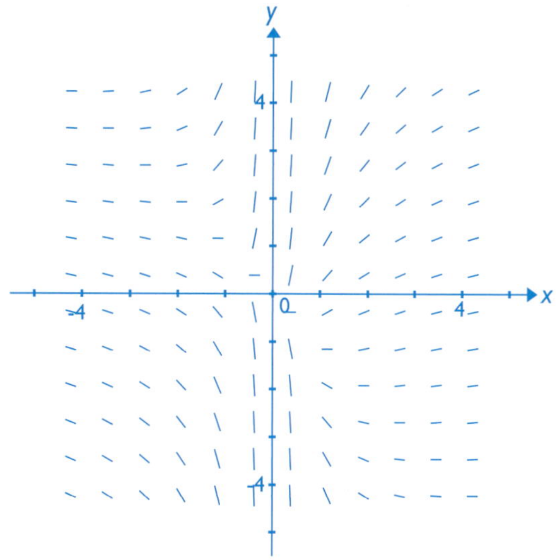

Figure 2.8

a. What is the slope of the tangent drawn at (4, 4)?

b. What is the slope of the tangent drawn at (−4, 0)?

c. Explain why the slope is not defined at the origin.

d. Describe the slopes of the solution curves along the x-axis.

7. Figures 2.9a–f show six slope fields. Match them with these differential equations.

i)	$y' = 2xy$	**ii)**	$y' = -xy$
iii)	$y' = xy$	**iv)**	$y' = 0.5xy + y$
v)	$y' = 0.5xy + x$	**vi)**	$y' = xy - x$

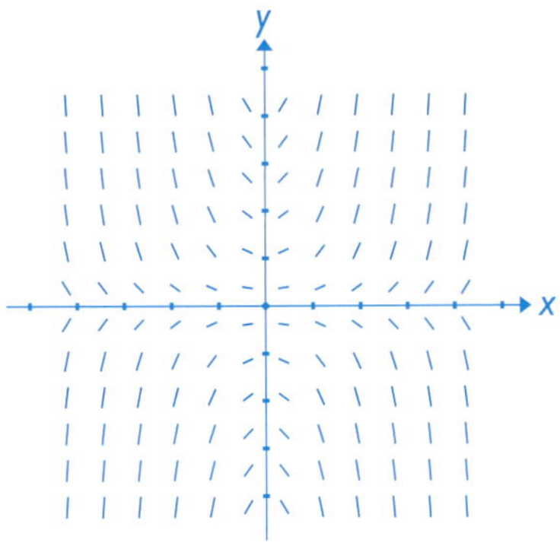

Figure 2.9a

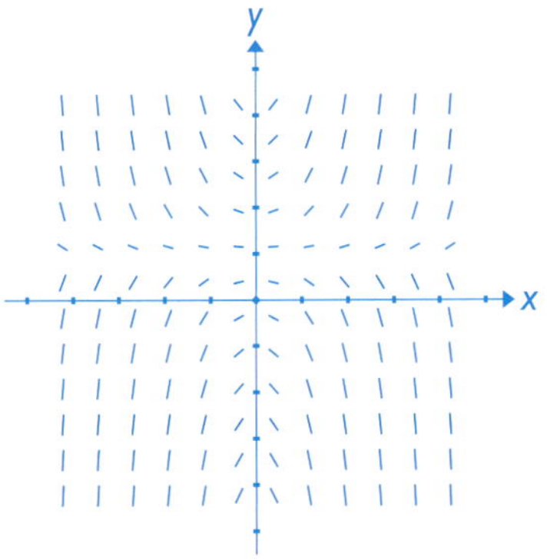

Figure 2.9b

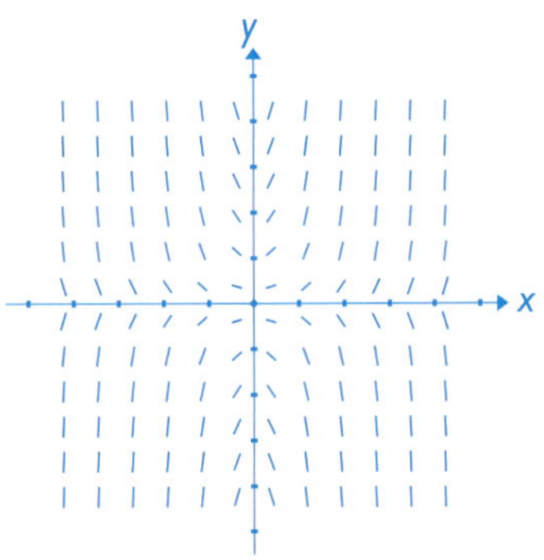

Figure 2.9c

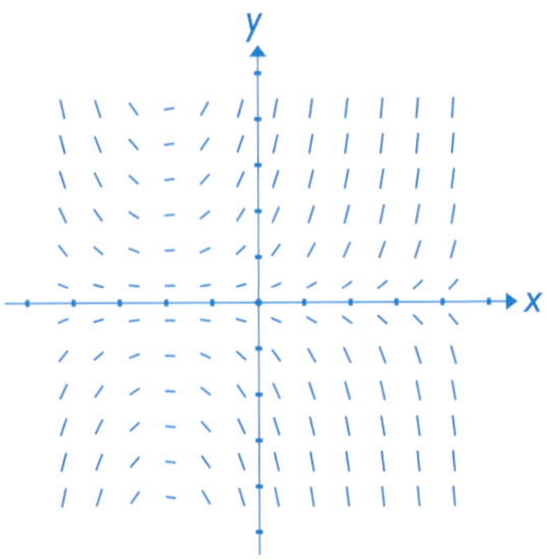

Figure 2.9d

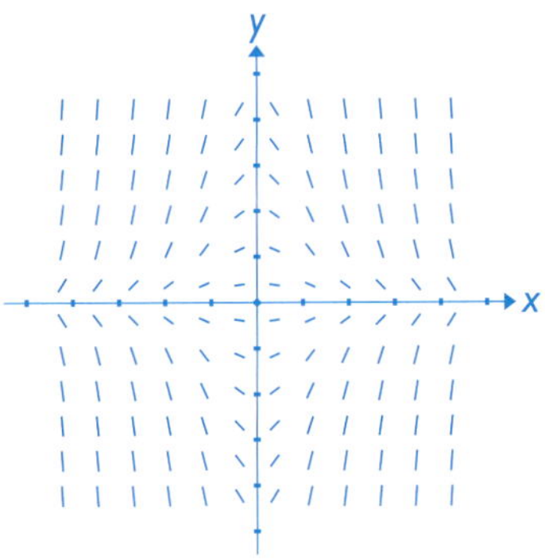

Figure 2.9e

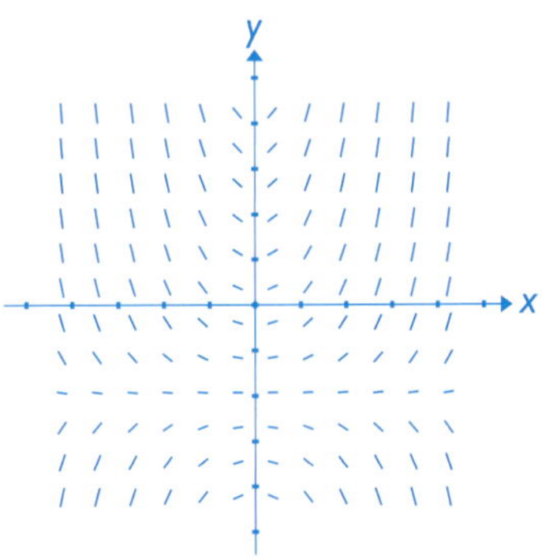

Figure 2.9f

8. Figures 2.10a–f show six slope fields. Match them with these differential equations.

i)	$y' = 0.5x$	**ii)**	$y' = -x$
iii)	$y' = -y$	**iv)**	$y' = 0.5y$
v)	$y' = 0.5y^2x$	**vi)**	$y' = -0.5y^2x$

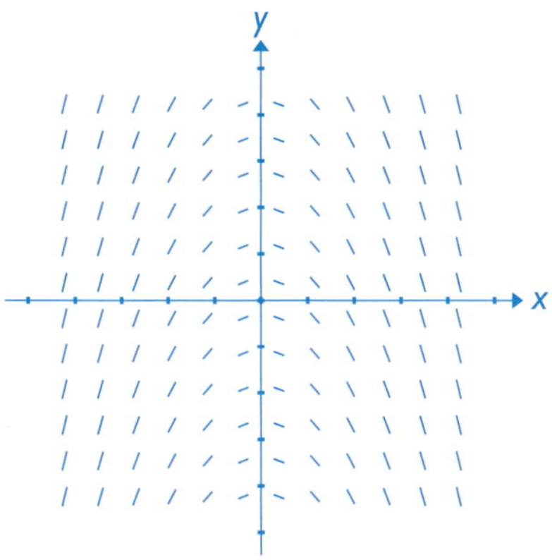

Figure 2.10a

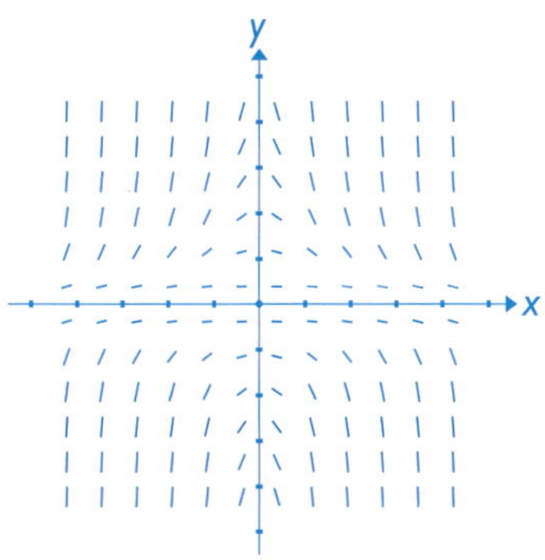

Figure 2.10b

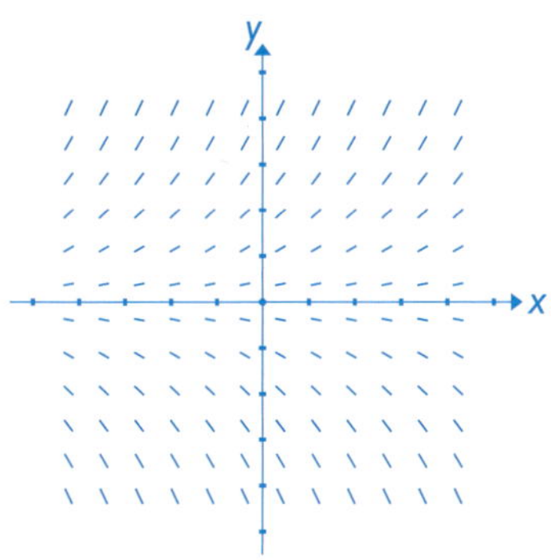

Figure 2.10c

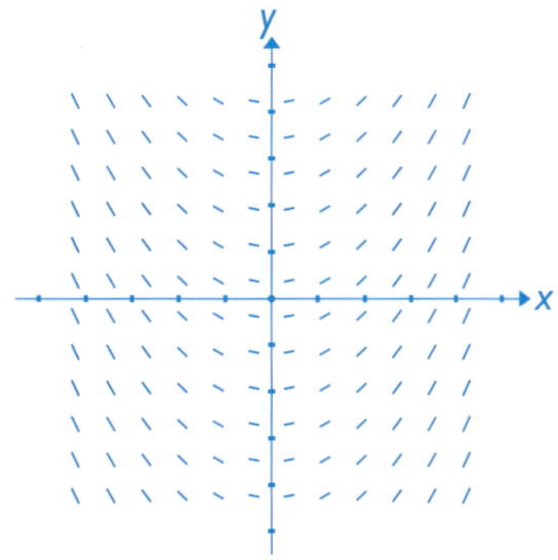

Figure 2.10d

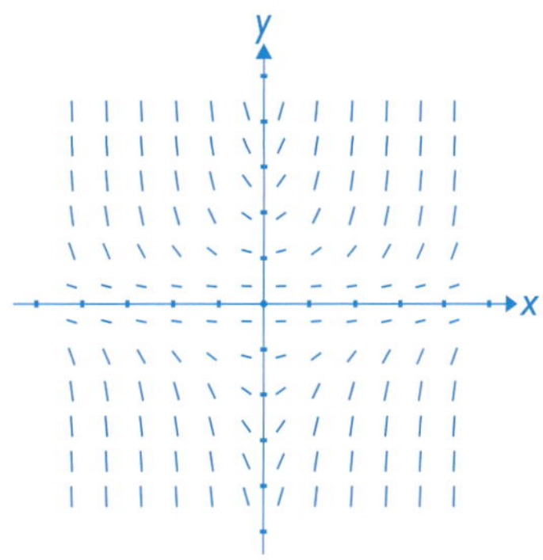

Figure 2.10e

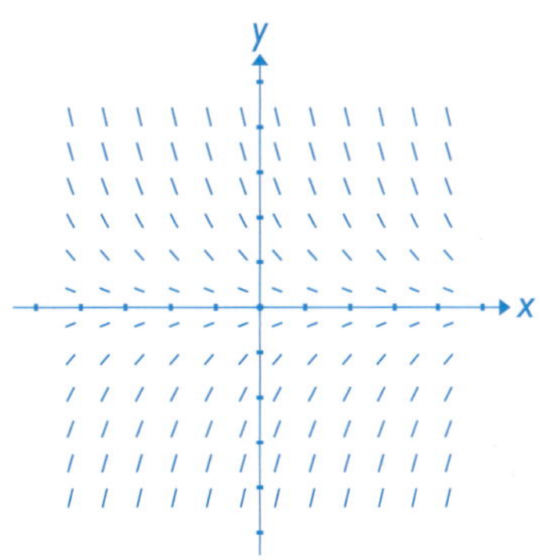

Figure 2.10f

9. Figures 2.11a–f show six slope fields. Match them with these differential equations.

i)	$y' = x + y$	**ii)**	$y' = x - y$
iii)	$y' = y - x$	**iv)**	$y' = 2x + 3y$
v)	$y' = 2y + 3x$	**vi)**	$y' = x + y - 8$

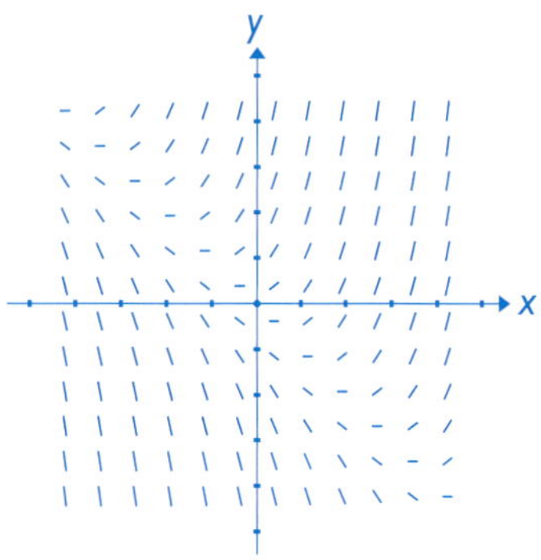

Figure 2.11a

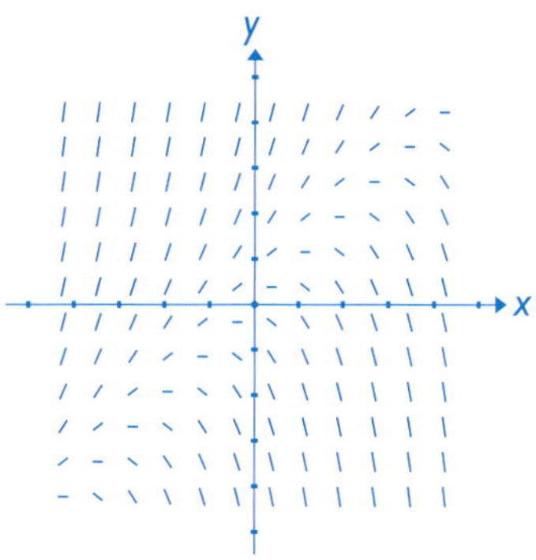

Figure 2.11b

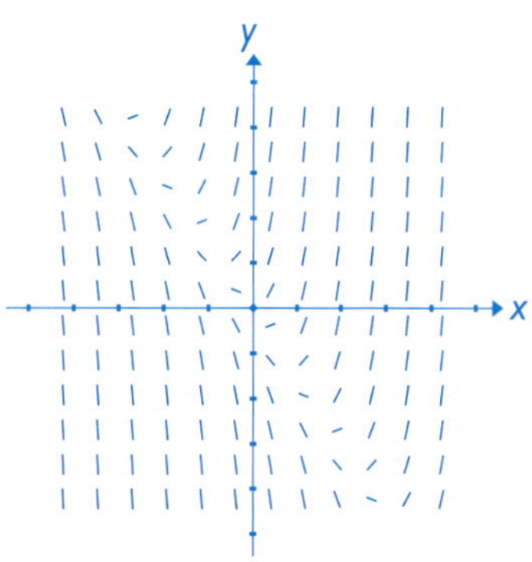

Figure 2.11c

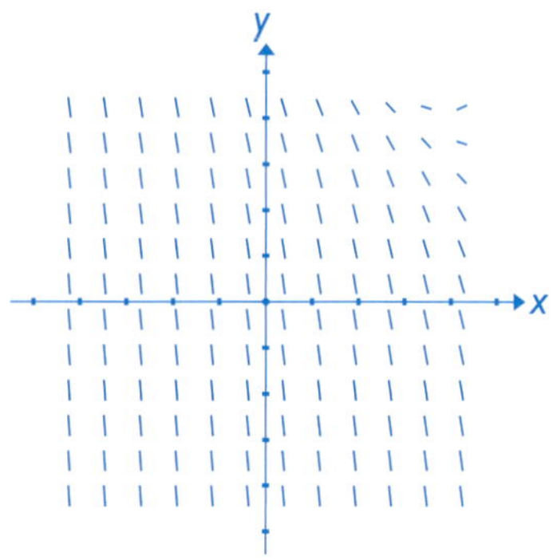

Figure 2.11d

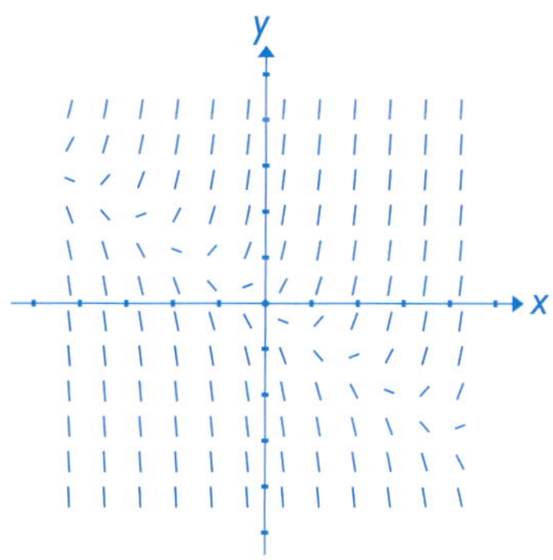

Figure 2.11e

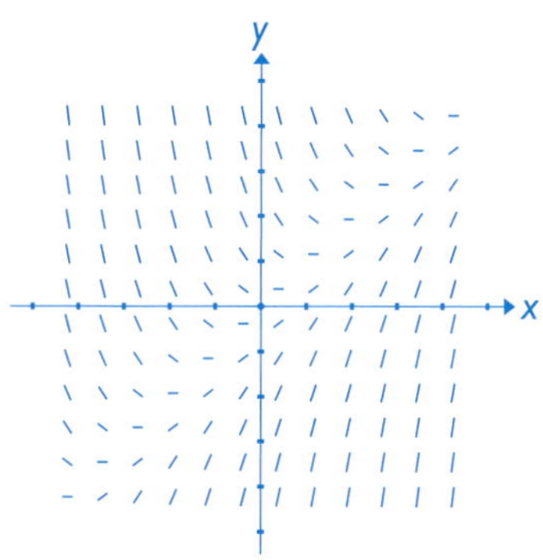

Figure 2.11f

10. Figures 2.12 a–f show six slope fields. Match them with these differential equations.

i) $y' = \frac{x}{y}$ **ii)** $y' = \frac{-x}{y}$

iii) $y' = \frac{-y}{x}$ **iv)** $y' = \frac{y}{x}$

v) $y' = \frac{y^2}{x^2}$ **vi)** $y' = \frac{x^2}{y^2}$

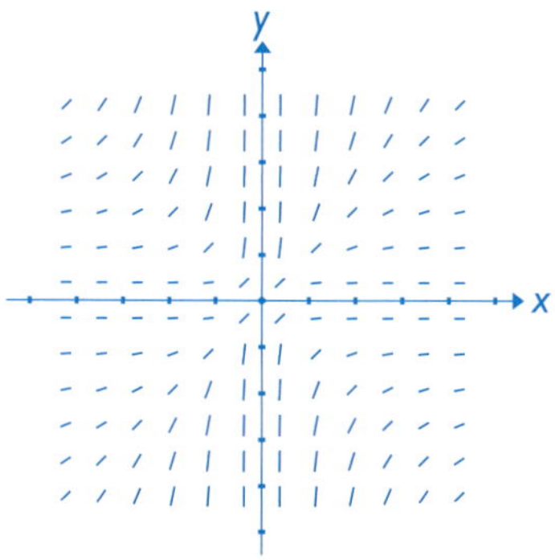

Figure 2.12a

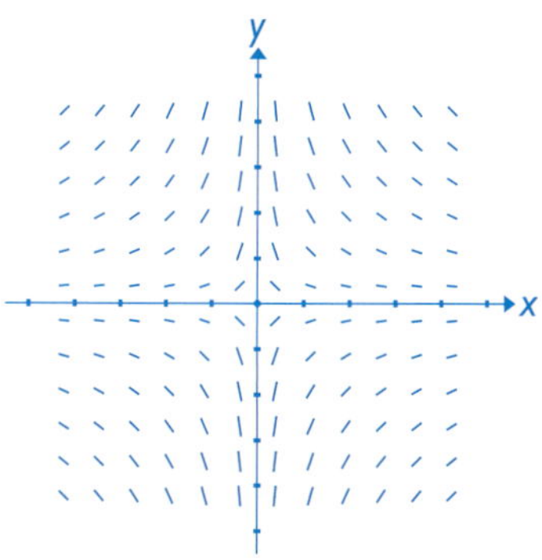

Figure 2.12b

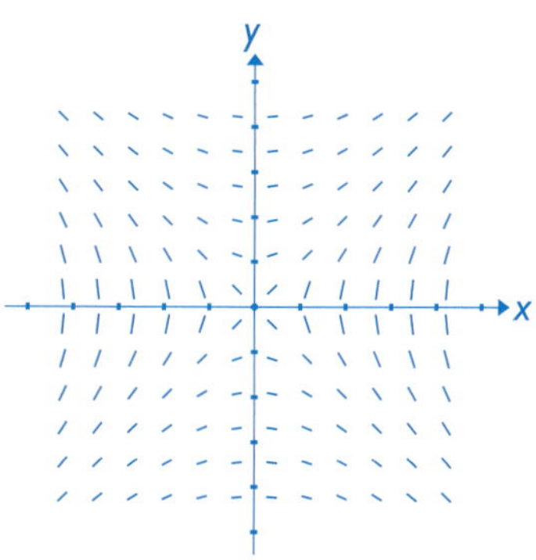

Figure 2.12c

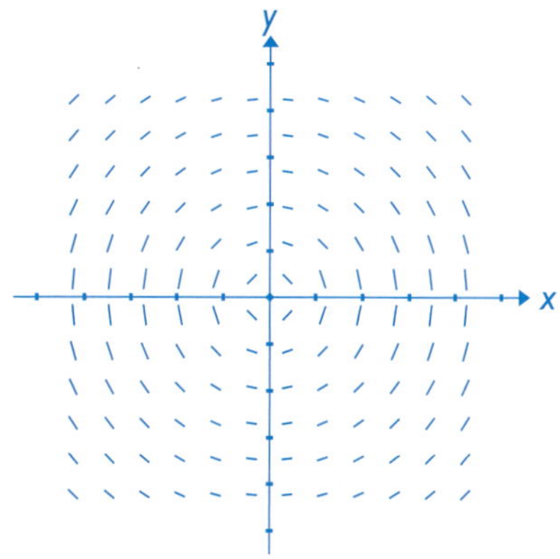

Figure 2.12d

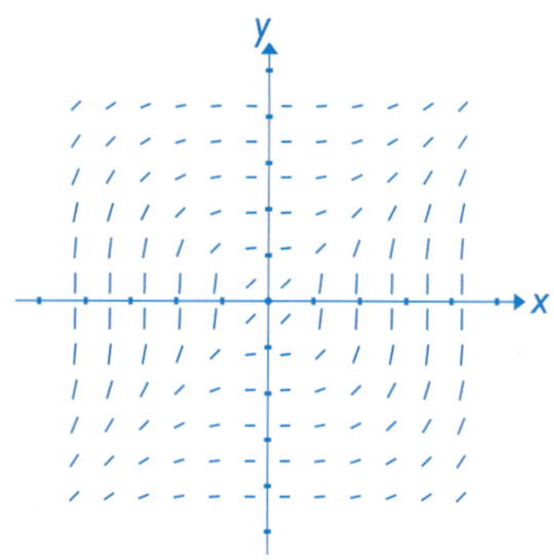

Figure 2.12e

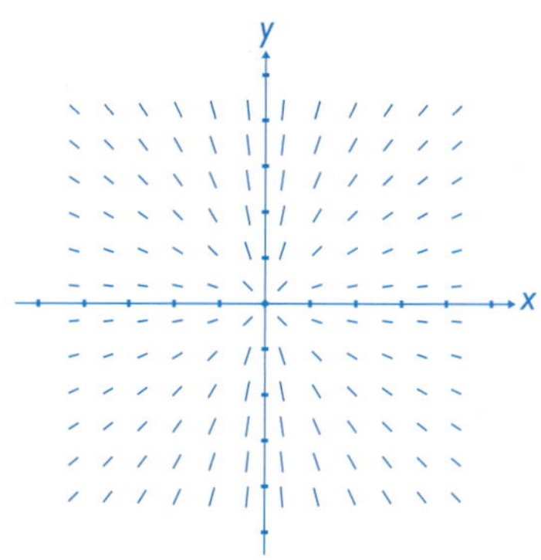

Figure 2.12f

11. Figure 2.13 shows a slope field for the differential equation $\frac{dy}{dx} = \frac{y - 2x}{y - x}$.

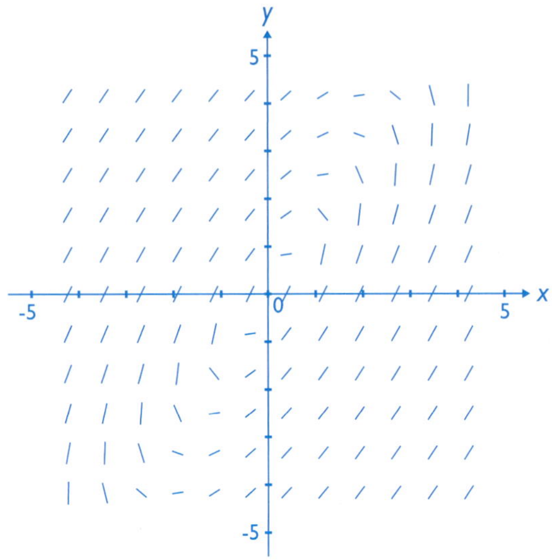

Figure 2.13

a. What are the slopes of the tangent lines along the diagonal $y = 3x$?

b. What kind of tangent lines appear along the diagonal $y = x$? Explain.

12. The differential equation $\frac{dy}{dx} = g(x)$ gives the derivative as a function of x. What information does the form of this derivative tell you about the slope field?

13. The differential equation $\frac{dy}{dx} = h(y)$ gives the derivative as a function of y. What information does the form of this derivative tell you about the slope field?

14. The differential equation $\frac{dy}{dx} = f(x^2)$ gives the derivative as a function of x^2. What information does the form of this derivative tell you about the slope field?

15. The slope field of a differential equation is shown in Figure 2.14. Is the derivative a function of x only, of y only, or of both x and y? Explain.

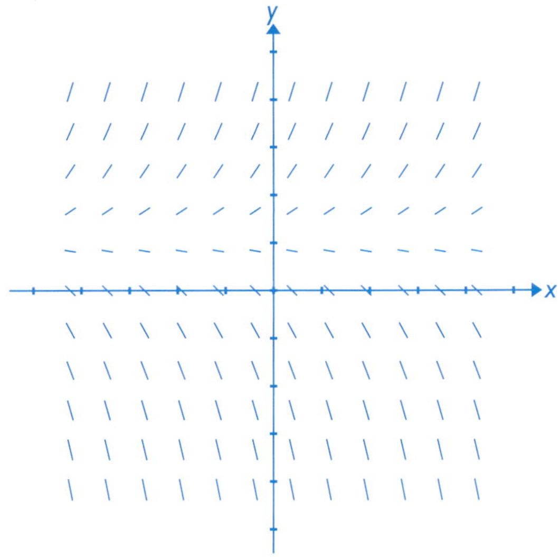

Figure 2.14

16. The slope field of a differential equation is shown in Figure 2.15. Is the derivative a function of x only, of y only, or of both x and y? Explain.

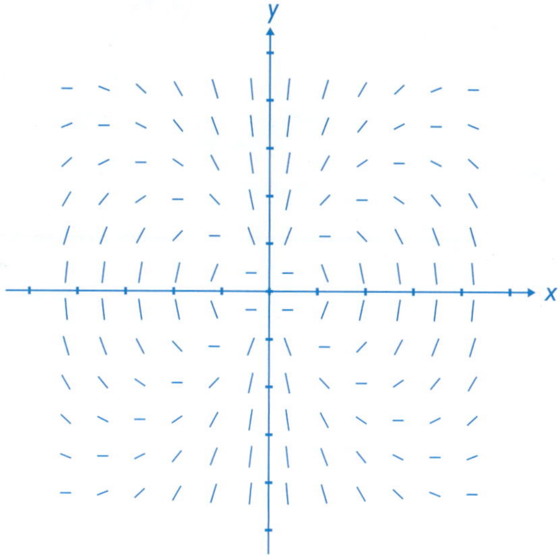

Figure 2.15

17. What differential equation might have the slope field shown in Figure 2.16?

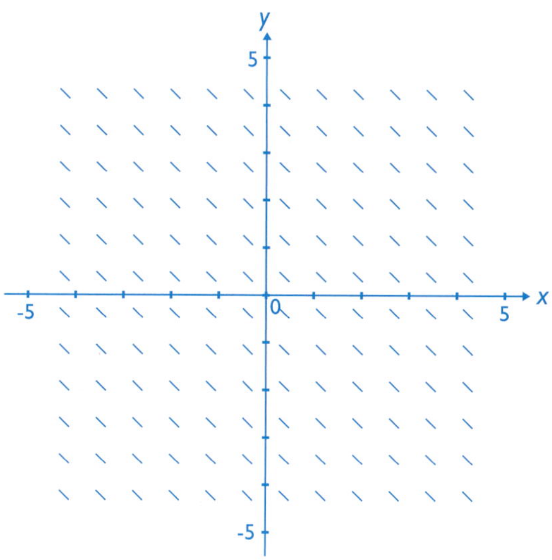

Figure 2.16

18. What differential equation has the slope field shown in Figure 2.17?

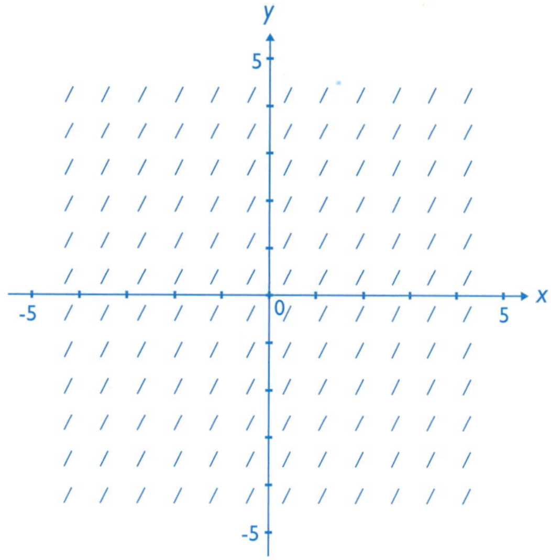

Figure 2.17

19. What kind of symmetry must slope fields of the differential equation $\frac{dy}{dx} = \frac{xy}{x^2 + y^2}$ have?

CHAPTER 3

Initial Value Problems and Particular Solutions

One of the most important applications of Calculus is to determine an unknown function, knowing only its derivative and one or more values of the function. Suppose, for example, a physicist in a 25°C room brews a pot of tea, starting with boiling water. Three minutes later the tea has cooled to 90°C. How long will it be before she can enjoy a cup of tea at 75°C?

We can model this problem using Newton's Law of Cooling, which holds that the rate at which an object cools is proportional to the difference between the temperature of the object and the surrounding temperature, which is assumed to be constant. For this situation Newton's Law may be expressed algebraically by the differential equation $\frac{dT}{dt} = k(T - 25)$, where $T(t)$ is the temperature of the tea at time t and k is the constant of proportionality. We also know that $T(0) = 100$, and $T(3) = 90$. In the problem set for this chapter you will have an opportunity to find a solution to this equation and to discover properties of the solution from its slope field.

A differential equation such as this, where one or more values of the function are known, is called an **Initial Value Problem** (or **IVP**) because the value of the function is often known at time $t = 0$. But the "initial" value can be at any point. The initial conditions allow us to determine the constants of integration in the general solution and so the solution to an initial value problem is just a particular solution of a differential equation. In this chapter we use slope fields to examine a number of different IVPs.

Example 1 Use a slope field to sketch the solution to $\frac{dy}{dx} = 3\sin(x)\cos(x)$ passing through (0, 1).

Solution The slope field is shown in Figure 3.1. The particular solution passing through (0, 1) is sketched by starting at the given point (0, 1) and drawing a curve that follows the tangent segments in both directions. Notice that the curve does not actually touch most of the tangents that are closest to the curve. Remember that a slope field is a graph indicating representative tangents. In general, there is no reason to assume that the particular tangent lines in a slope field will be the tangent lines of any particular solution.

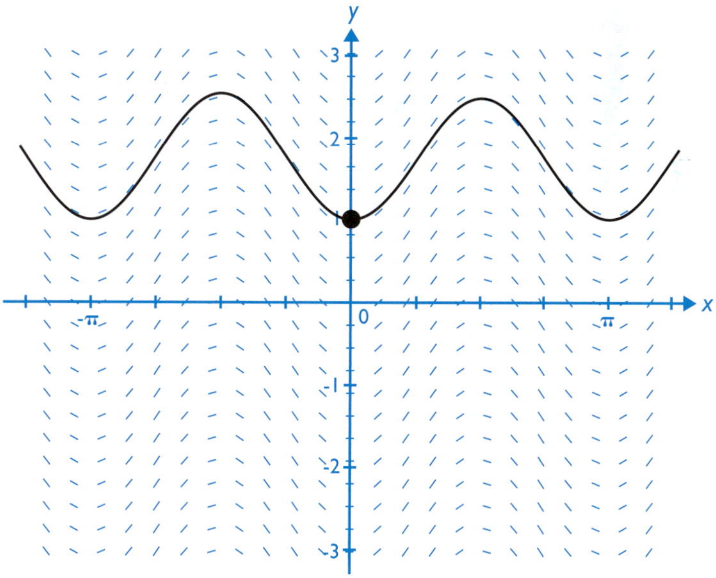

Figure 3.1

The solution appears to be an inverted cosine wave with period π and amplitude $\frac{3}{4}$. We can confirm this by antidifferentiation. The substitution $u = \sin(x)$ transforms the problem into $y = \int 3\sin(x)\cos(x)dx = 3\int u\,du$. So $y = \frac{3\sin^2 x}{2} + C$. Since the solution passes through (0, 1), $C = 1$ and $y = \frac{3\sin^2 x}{2} + 1$.

This particular solution does not look like a cosine equation, but the double-angle formula $\cos(2x) = 1 - 2\sin^2(x)$ lets us rewrite the solution as:

$$y = \frac{3[1 - \cos(2x)]}{4} + 1 = \frac{7}{4} - \frac{3}{4}\cos(2x).$$

CAUTION

It is easy to be misled by a slope field. Realize that any particular solution you graph may not have points in common with the nearest tangents given in the slope field. Figures 3.2a and 3.2b, where the solution curve $y = x^3$ is shown on two different slope fields, illustrate this fact. In Figure 3.2a there are only 25 tangent segments drawn and it is clear that none of them is exactly tangent to this curve. Nevertheless, if you were asked to sketch the solution passing through the origin, it is this curve that you should draw.

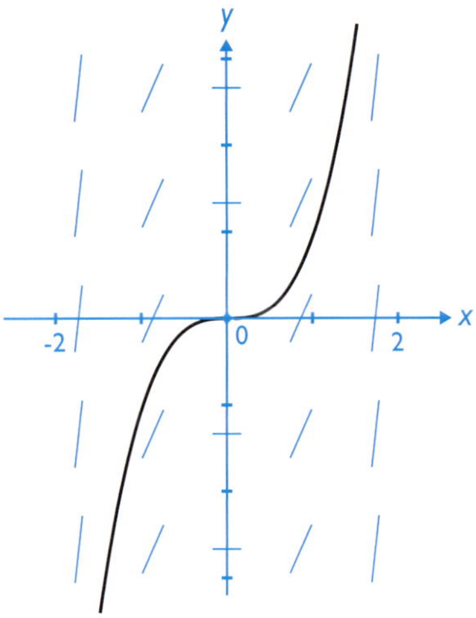

Figure 3.2a

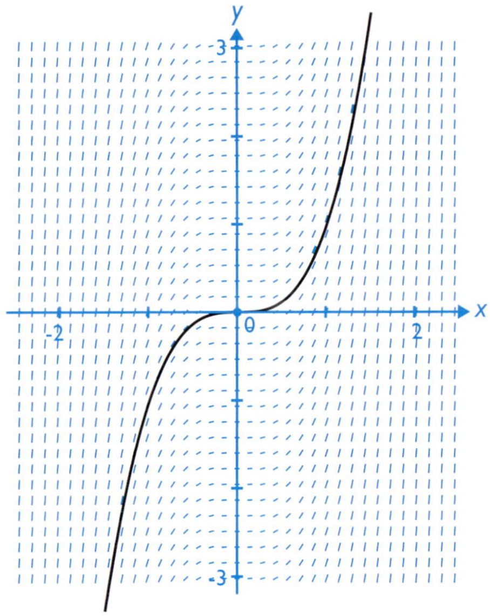

Figure 3.2b

Figure 3.2b shows a slope field for the same differential equation, but this slope field is made by representing 1225 tangent lines (35 by 35). It is still the case that $y = x^3$ does not touch any of these tangents, but the segments are so short and close that the error made in assuming tangency is slight.

The next chapter is devoted to understanding the peculiarities and pitfalls of using slope field graphs. For now, be cautious when drawing particular solutions. Do not try to make the curves touch the line segments you see. Instead, just try to follow their general contour.

Example 2 Use a slope field to sketch the solution to the initial value problem $\frac{dy}{dx} = \frac{1}{y}$ passing through $(-2, -1)$.

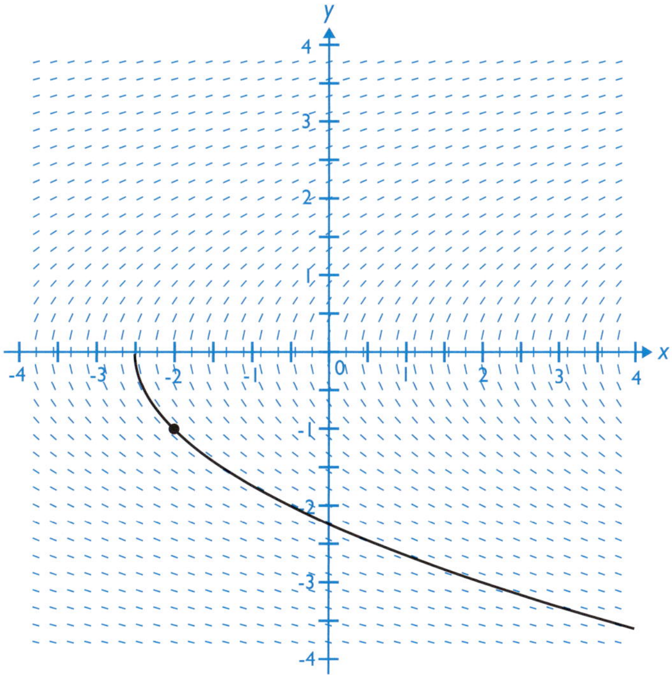

Figure 3.3

Solution The solution is sketched in Figure 3.3. The slope field appears to be composed of sideways parabolas but the particular solution through $(-2, -1)$ stops at the x-axis. Why should that be?

Since the derivative is not defined on the x-axis, where $y = 0$, we cannot extend the curve through or past this boundary. In fact, there are two *types* of solutions to this differential equation: each is half of a sideways parabola. If the initial value is above the x-axis then the solution will be entirely above the x-axis. If the initial value is below the x-axis then the solution will be entirely below the x-axis.

In working this example and others in the problem set, you may be greatly tempted to continue a graph across a vertical tangent line. But you must stop yourself. Situations giving rise to this type of error will be explored in great detail in Chapter 4.

Example 3 Use a slope field to sketch the solution to $\frac{dy}{dx} = \frac{x + y}{x - y}$ passing through (3, 2).

Solution Figure 3.4 shows a slope field for this differential equation with the solution curve through (3, 2). The field looks to be made of spirals around the origin. But why does this solution seem to be restricted to just part of a loop?

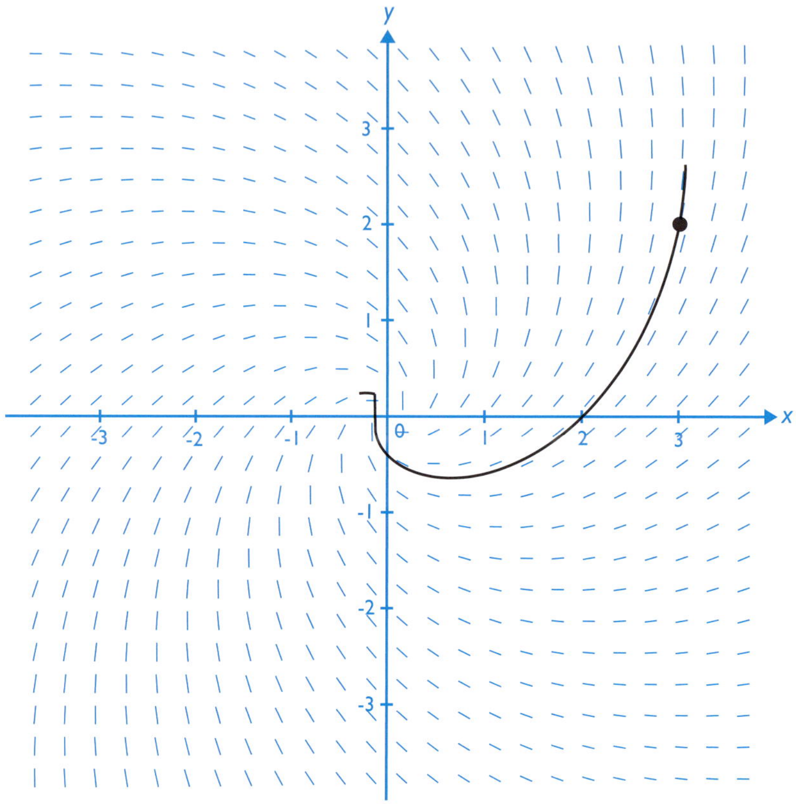

Figure 3.4

The derivative does not exist at (0, 0) or (3, 3), since those values make the denominator zero. In fact, the derivative does not exist at any point along the line $y = x$. Therefore, when sketching the solution curve on the slope field we do not extend the curve across the line $y = x$. Some solutions will lie below $y = x$, while others will lie above it. The line $y = x$ acts as a boundary, separating different types, or categories, of solution curves. While our eye may put together spirals, the actual solutions are segments of these spirals, some above the boundary line and some below.

Example 4 Use a slope field for $\frac{dy}{dx} = x + y$ to visualize solutions passing through (2, 0), (0, −3), and (−1, 0). Explain the peculiar behavior of the solution through (−1, 0).

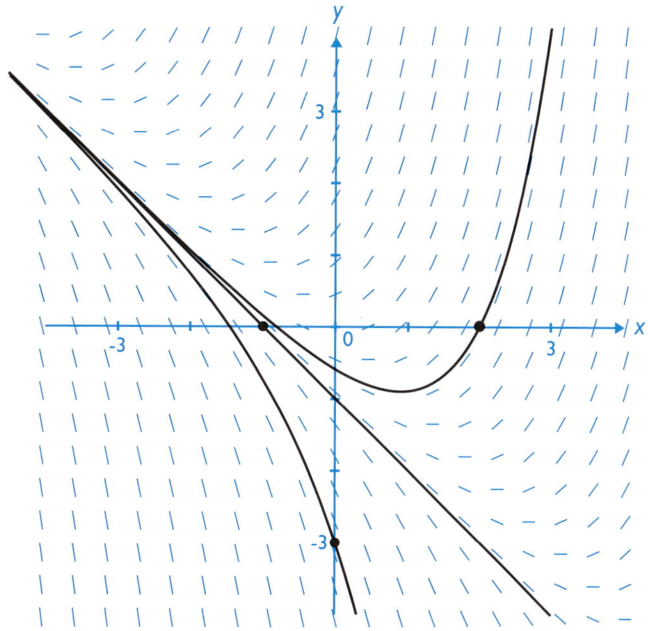

Figure 3.5

Solution The solutions to all three initial value problems are shown in Figure 3.5. While we do not know how to solve the equation algebraically, it is easy to confirm that the line $y = -1 - x$ is a solution passing through (−1, 0). Since $x + y = -1$, the differential equation reduces to $\frac{dy}{dx} = -1$. The solution is a line whose slope is −1 and the initial value (−1, 0) gives the particular line shown in the figure.

Since any point on this line could serve as an initial value for the same problem, no other solution curve can intersect this line. If another curve did appear to cross, then the crossing point would serve as a different initial value, but the solution to that IVP would be the $y = -1 - x$ line, not the other curve. Therefore, solutions passing through points where $y > -x - 1$ will lie entirely above this line, while those passing through points below this line will lie entirely below.

This analysis points out an important property of solutions to differential equations worth stating by itself:

> Different solutions to the same differential equation cannot intersect.

Something else worth noting is suggested in Figure 3.5. It appears that the two non-linear solution curves are asymptotic to the linear solution $y = -1 - x$. Why should that be so? You will see similar behavior of solution curves in the problem set that follows and, in Problem 7, have the opportunity to develop part of the answer to this question.

Problems

www.peoplescollegeprep.com/slopefields.html

1. Figure 3.6 shows a slope field for $t\frac{dz}{dt} = t^2 - \frac{dz}{dt}$ with the particular solution through (3, 0) graphed.

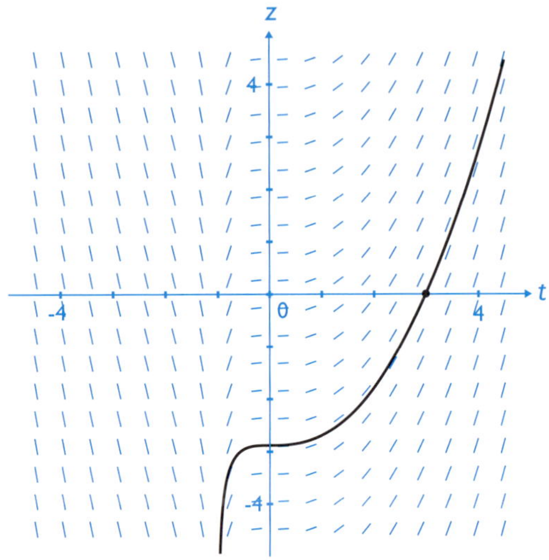

Figure 3.6

a. On a copy of the slope field, sketch the particular solution passing through (0, 2). Use your sketch to approximate $z(2)$.

b. Solve this IVP algebraically and graph your solution. Compute the exact value of $z(2)$ that you approximated in **1a**.

2. Figure 3.7 shows a slope field for $\frac{d\theta}{dt} = \sqrt{\theta + 1}$.

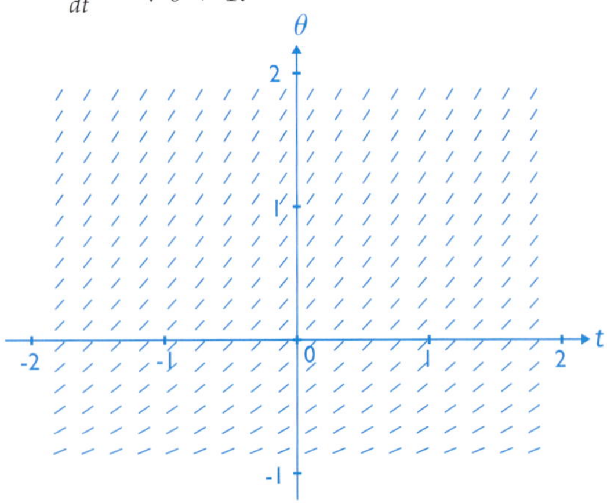

Figure 3.7

a. On a copy of the slope field, sketch the particular solution passing through the origin and use it to approximate $\theta(1)$.

b. Solve this IVP algebraically and graph your solution. Compute the exact value of $\theta(1)$ that you approximated in **2a**.

3. Figure 3.8 shows a slope field for the differential equation $\frac{dy}{dx} = \sin^3(x)\cos(x)$.

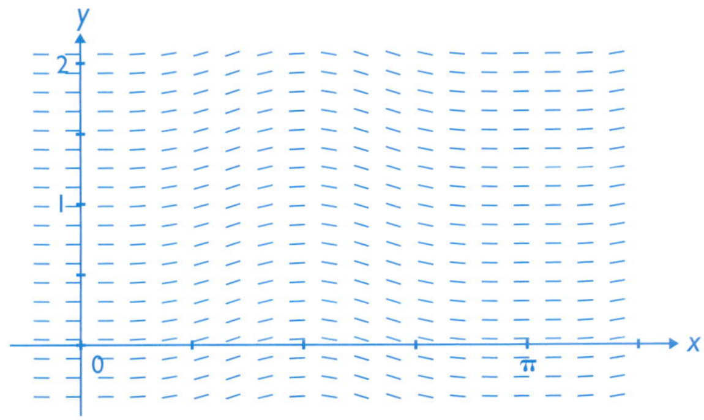

Figure 3.8

a. On a copy of the slope field, sketch the particular solution passing through $(\pi, 1)$ and use it to approximate $y(\frac{\pi}{2})$.

b. Solve this IVP algebraically and graph your solution. Compute the exact value of $y(\frac{\pi}{2})$ that you approximated in **3a**.

4. Figure 3.9 shows a slope field for $\frac{dy}{dx} = \sec^2(x)\sqrt{\tan(x)}$.

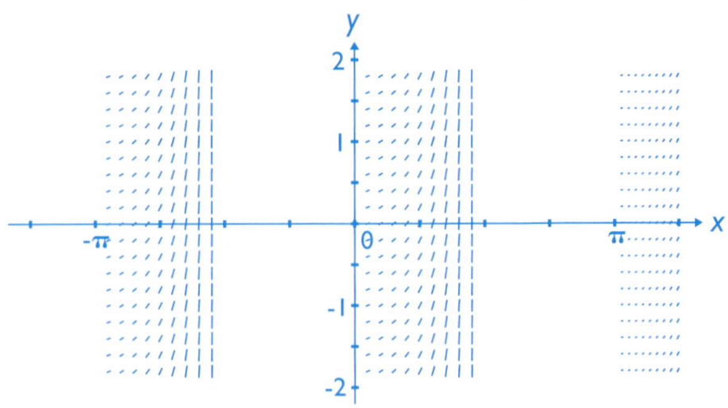

Figure 3.9

a. On a copy of the slope field, sketch the particular solution passing through (0, 0) and use it to approximate $y(\frac{\pi}{4})$.

b. Solve this IVP algebraically and graph your solution. Compute the exact value of $y(\frac{\pi}{4})$ and compare it to the approximate value you computed in **4a**.

c. This slope field looks strange, its tangent segments appearing in alternating bands. What is the exact width of each band? Explain.

5. The slope field in Figure 3.10 shows an approximate solution to $\frac{dy}{dx} = x - y^2 + 1$ passing through the origin.

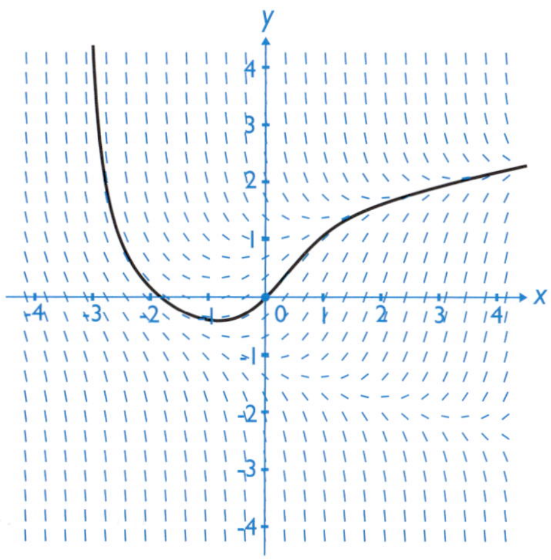

Figure 3.10

a. On a copy of Figure 3.10 sketch the solution passing through $(1, -2)$ and use it to approximate $y(-2)$.

b. On a copy of Figure 3.10 sketch the solution passing through $(1, 3)$ and use it to approximate $y(3)$.

6. Figure 3.11 is a slope field for the differential equation $\frac{dy}{dx} = (y - 2)(y + 1)$. It has two boundary lines and separates the solutions into three categories.

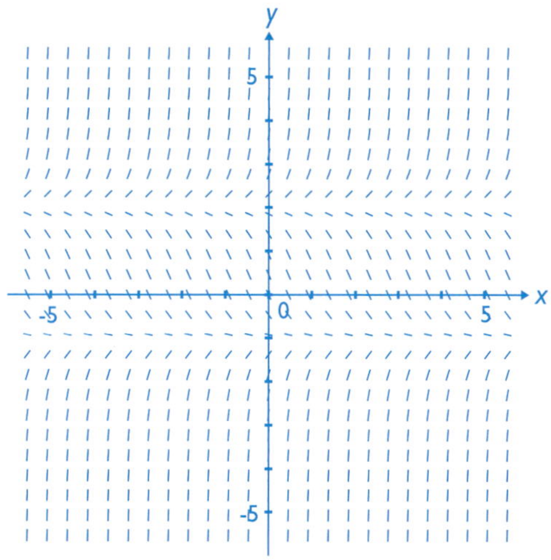

Figure 3.11

a. Find the equations of the boundary lines and explain why they are boundaries.

b. Sketch the solution passing through $(2, 0)$ and the solution passing through $(0, -3)$.

7. The slope field in Figure 3.5 for Example 4 suggested that the line $y = -1 - x$ was an asymptote to all other solutions to the differential equation $\frac{dy}{dx} = x + y$. In this problem we develop part of the answer by looking more closely at that slope field.

a. Show that $y = -1 - x$ is a solution to this differential equation.

b. Show that there is no other linear solution to this differential equation. (Assume that $y = mx + b$ is a solution and show that $m = b = -1$.)

c. Choose any point (a, b) below the line $y = -1 - x$ and consider the particular solution passing through (a, b). First show that the slope at (a, b) is negative. Use that fact to explain why a point (x, y) on the solution curve with $x < a$ must have $y > b$. Explain why that shows that the (x, y) must be closer to the line $y = -1 - x$ than (a, b).

d. Explain why your answers to the above questions imply that $y = -1 - x$ is an asymptote to every solution curve that lies below the boundary line.

8. Figure 3.12 is a slope field for the differential equation $\frac{dy}{dx} = 2x - y + 1$.

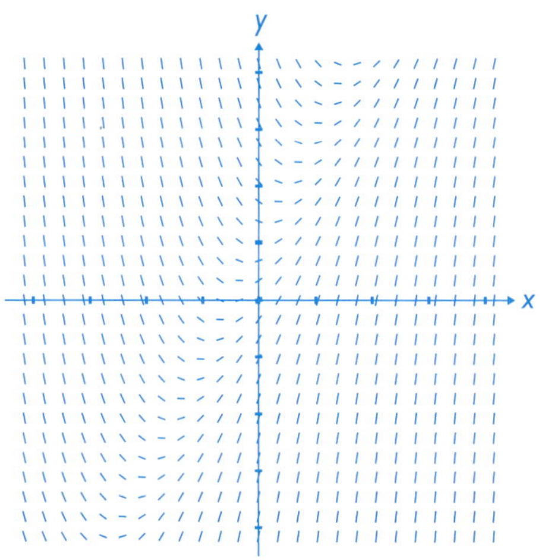

Figure 3.12

It appears to have a boundary line. Find the equation of this boundary by assuming that y is a linear function of x. That is, assume that $y = mx + b$ and substitute this into the differential equation and solve for m and b.

9. Figure 3.13 is a slope field for the differential equation $\frac{dy}{dx} = x + y - 6$. It appears to have a boundary line.

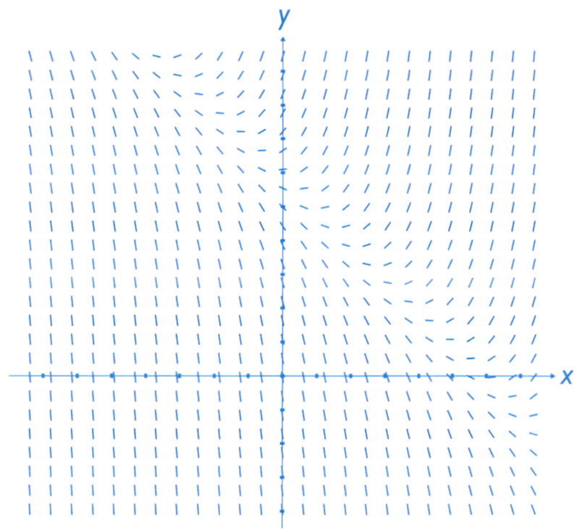

Figure 3.13

a. The boundary might seem like $x + y = 6$, but it is not. How can you be certain that it is not $x + y = 6$?

b. Find the equation of this boundary line by assuming that y is a linear function. (Substitute $y = mx + b$ into the differential equation and solve for m and b.)

c. Explain why there are two types of solutions.

10. Figure 3.14 shows a slope field for some differential equation.

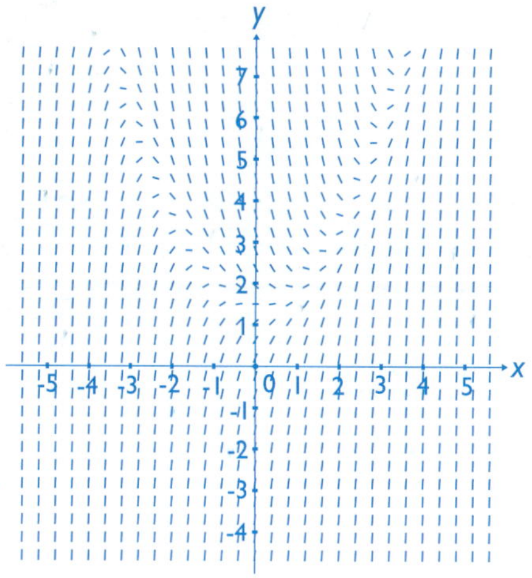

Figure 3.14

a. On what interval, approximately, is the solution through $(0, -4)$ increasing?

b. On what interval, approximately, is the solution through $(-2, 0)$ concave down?

11. Figure 3.15 shows a slope field for some differential equation.

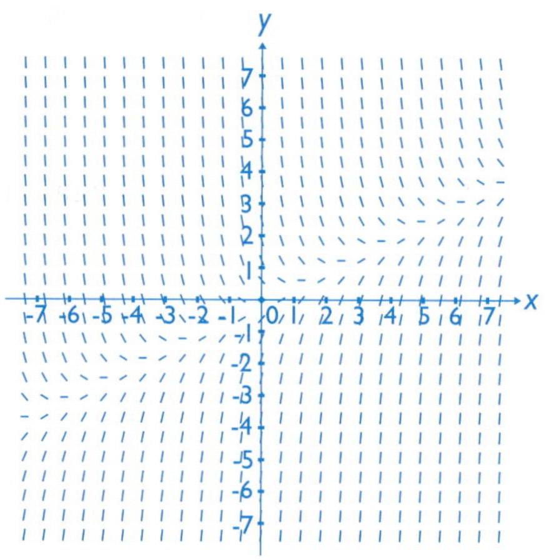

Figure 3.15

a. On what interval, approximately, is the solution through $(2, 0)$ increasing?

b. On what domain, approximately, are the solutions concave down?

12. Figure 3.16 shows the solution to the differential equation $\frac{dT}{dt} = -0.0477(T - 25)$ with $T(0) = 100$.

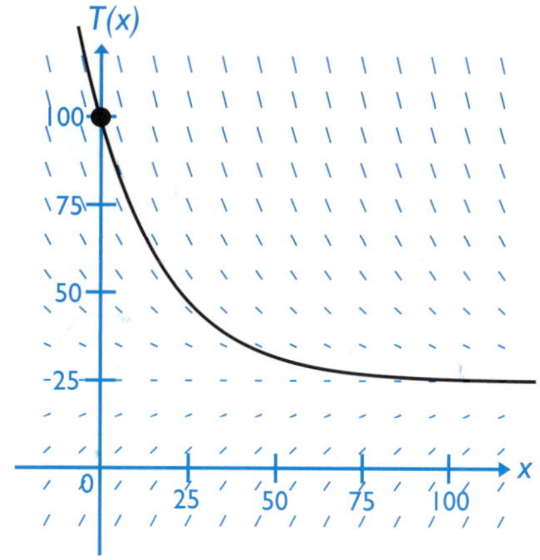

Figure 3.16

a. Solve this equation algebraically and show that $T(3) = 90$. That is, show that this is the same problem as the one discussed in the introduction to this chapter.

b. The solution curve seems to be leveling off to a horizontal asymptote. What is the equation of the asymptote? What is the meaning of this asymptote in the context of the tea cooling problem?

c. What feature of the slope field could have helped you to predict that the solution would have a horizontal asymptote?

CHAPTER 4

Avoiding Pitfalls in Slope Fields

A slope field can be a powerful aid to understanding the solutions to a differential equation; however, as with any tool, there is the possibility of misapplying it. A slope field can be quite confusing when the domain of the derivative is in some way restricted so that there are locations where the derivative fails to exist. We have seen several examples of such slope fields in previous chapters, but now we will look more deeply to see if we can clear up some of the confusions. The first example returns to a slope field we have seen before.

Example 1 What types of solutions to the differential equation $\frac{dy}{dx} = \frac{1}{y}$ are suggested by the slope field in Figure 4.1? Sketch the solution passing through $(-2, -1)$.

Solution The solutions seem at first to be one family of sideways parabolas. And if not careful, we may mistakenly sketch the solution through $(-2, -1)$ as shown in Figure 4.2. This curve is *incorrect* because it is not the graph of a function. Only the bottom half of the parabola should be drawn as part of the solution curve.

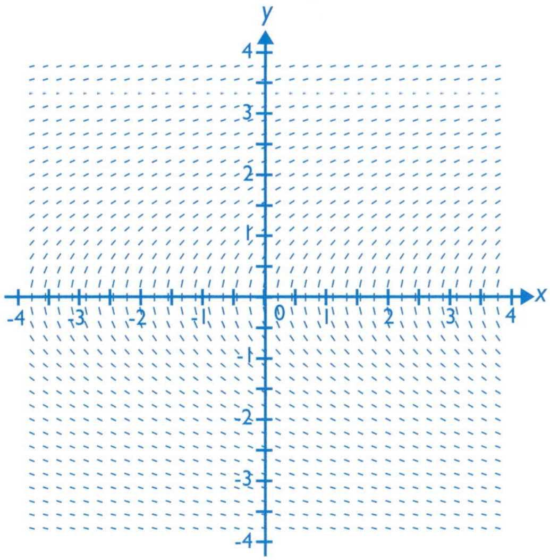

Figure 4.1

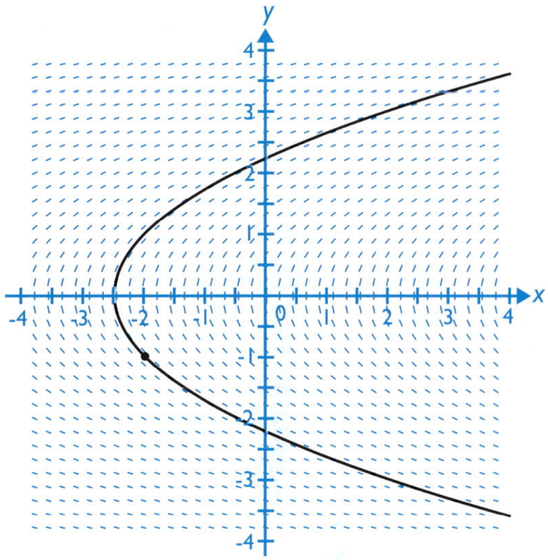

Figure 4.2

The parabolas are so obvious in the slope field that the urge to draw the complete curve is almost overwhelming, but we must resist that urge. There are two clues to help us find the correct solution—one in the algebraic form of the equation as a fraction, and one in the slope field. First, note that the denominator of the equation of the derivative is zero when $y = 0$. This is a warning to treat the x-axis (where $y = 0$) with special attention. Second, the sideways parabolas that we see in the slope field are not functions. As soon as we begin drawing a curve that fails the **vertical line test**, we should stop and reconsider our work.

The correct interpretation of the slope field, therefore, is that there are two different families of solution curves, the top halves and the bottom halves of parabolas. Solutions in the range $y > 0$ are all concave down and increasing, while those in the range $y < 0$ are concave up and decreasing. The two families separate at the x-axis, which contains vertical tangents at the turning points of the parabolas. The x-axis acts as a boundary line separating one family of solutions from the other.

Example 1 illustrates the caution we must take when a slope field indicates vertical tangent lines. But the pitfalls in slope fields can be even more difficult to see, as the next example illustrates.

Example 2 Find the general solution to the differential equation $\frac{dy}{dx} = \frac{1}{3y^2}$. Then use the slope field in Figure 4.3 to sketch the solution passing through (1, 1).

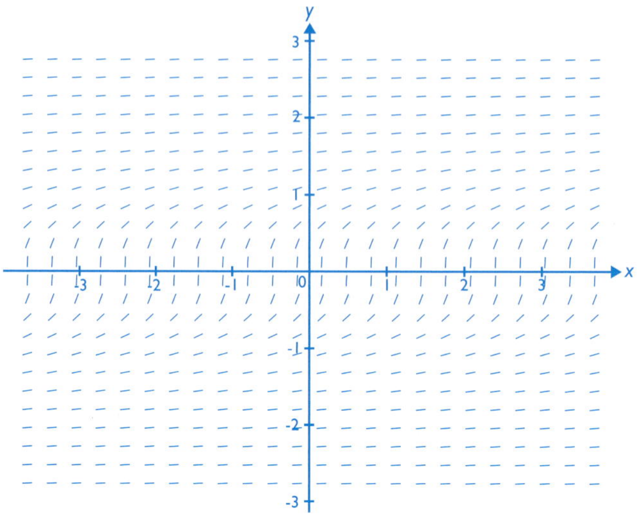

Figure 4.3

Solution Begin by separating the variables to get $3y^2dy = dx$ and then antidifferentiate both sides. This gives the general solution $y^3 + C = x$, the graph of a sideways cubic or, if we solve it for y, the graph of a cube root function, $y = \sqrt[3]{x - C}$.

Unlike the sideways parabolas of the previous example, these cube root equations are functions on any domain, so the temptation is to graph the complete function. However, this would be just as much a mistake as previously. Remember that we are working from a derivative that fails to exist when $y = 0$ and where the slope field shows what appears to be a vertical tangent line. Moreover, although this curve does not turn back upon itself at the vertical tangents, it is still incorrect to continue the curve anywhere *past* this point of discontinuity. The correct solution is shown in Figure 4.4—the top half only of a cube root function.

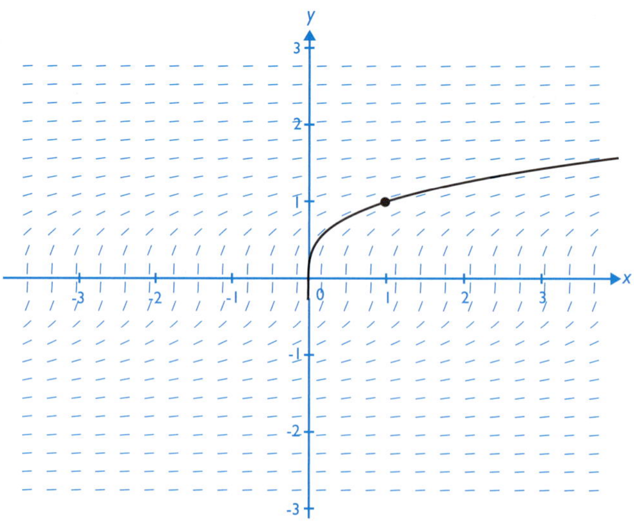

Figure 4.4

To see why it would be wrong to include the other half of this cube root, consider the discontinuous cubic relation $x = \begin{cases} y^3 & y > 0 \\ y^3 - 1 & y \leq 0 \end{cases}$ that passes through (1, 1) and is graphed in Figure 4.5. Differentiate implicitly to find that the derivative is $\frac{dy}{dx} = \frac{1}{3y^2}$ for $y \neq 0$. This is the same derivative that we just considered in Example 2. The graph of this piecewise function includes two disconnected curves.

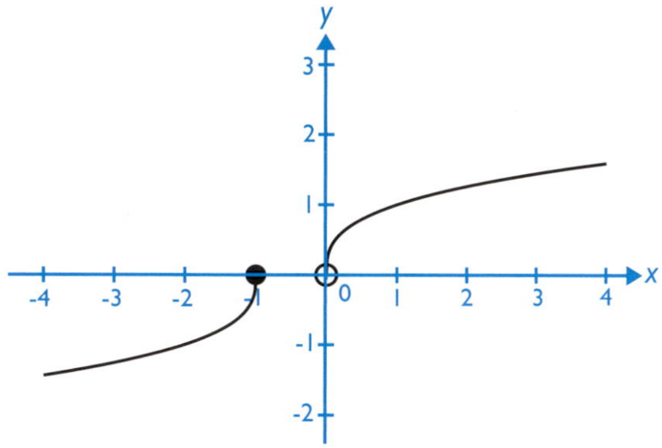

Figure 4.5

We can find these same curves in the slope field in Figure 4.3. Because this function satisfies the conditions of the differential equation, it is just as good a solution as the continuous cube root curve we were tempted to draw in Example 2. Because these are two different possible solutions to the same differential equation, we cannot presume to know how to extend the curve in Figure 4.4 below the x-axis. Without additional information we should draw only the top half.

The Problem of Holes

The previous discussion shows that curves with vertical tangents (where the derivative becomes infinite) create a problematic situation in a slope field. Derivatives that fail to exist because they have one or more holes present a similar difficulty. The following example will illustrate the point.

Example 3 Sketch the solution to $f'(x) = \frac{x^2 - 4}{x - 2}$ passing through (0, 0).

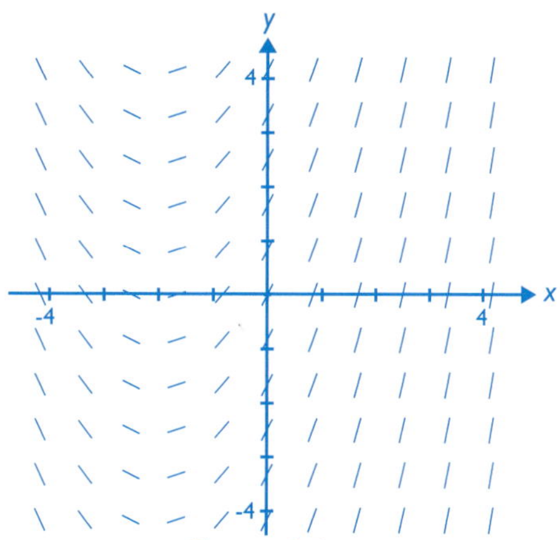

Figure 4.6a

Solution The slope field in Figure 4.6a suggests that the solution is a parabola. The solution *is* a parabola, but if we do not notice the hole at $x = 2$, we will likely draw the wrong particular solution because this slope field does not identify the problem at $x = 2$. Figure 4.6b redraws the slope field in a different window. This image emphasizes the gap at $x = 2$ and so we are more likely to perceive the correct particular solution, restricted to the domain $x < 2$. As in the previous example, we cannot extend our solution past the location where the derivative fails to exist.

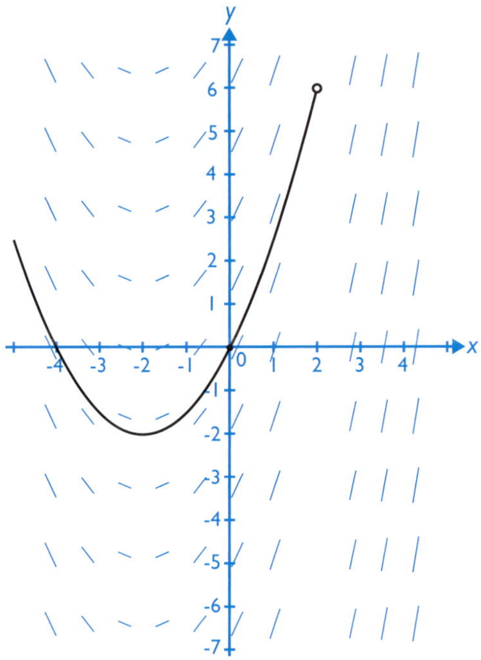

Figure 4.6b

This example raises a new caution, one that you have probably seen in other contexts. Calculators and computers are wonderful tools, but they have limitations. They will often be unable to show the location of holes. Because such a hole may radically affect the domain of the solution we must look to the equation itself for guidance and not just the slope field.

Examples 2 and 3 were chosen to make clear why it would be a mistake to extend a solution curve through a hole or past a vertical tangent line. The next example illustrates another possible confusion, because the derivative has neither a hole nor a vertical tangent line.

Example 4 Use the slope field in Figure 4.7a to sketch the solution curve passing through $(3, -1)$.

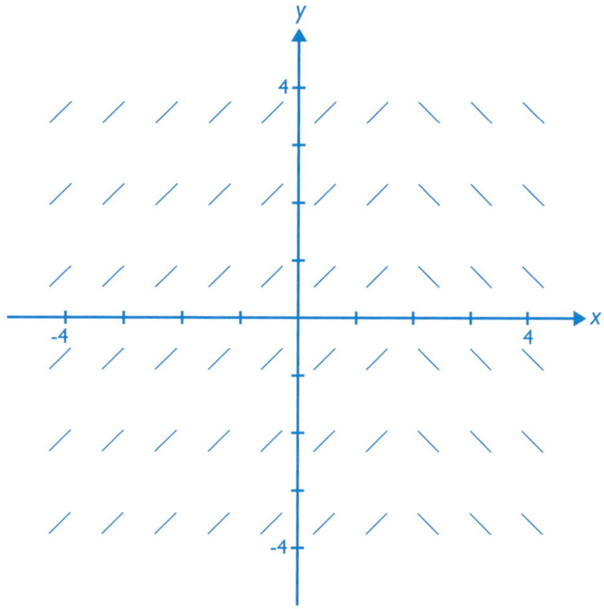

Figure 4.7a

Solution The slope field suggests that the general solution is a family of absolute value functions $y = C - |x - 2|$ and that the particular solution through $(3, -1)$ is, therefore, $y = |x - 2|$. However, this is not correct. The difficulty, as before, is that the derivatives of these absolute-value functions do not exist at certain locations—namely, at the corners.

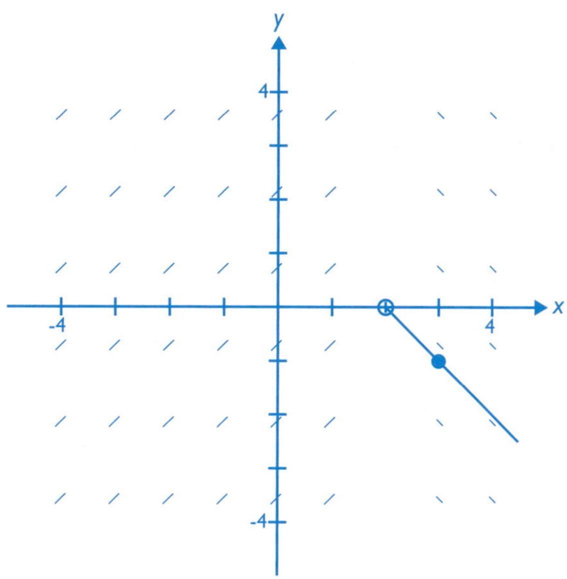

Figure 4.7b

Figure 4.7b shows another slope field for this function, drawn in the same window but with the tangent segments spaced exactly 1 unit apart. This new window clearly shows that there is a problem along $x = 2$. Because there are no tangent lines there, particular solutions should not cross $x = 2$. Remember that even a continuous function will fail to have a derivative at a corner or a cusp. Because the derivative does not exist when $x = 2$, the solution passing through $(3, -1)$ is the right-hand branch only of the absolute value function $y = -|x - 2|$, or $y = 2 - x$ for $x > 2$.

As in Example 3, the slope field has a gap where the derivative is undefined, but in this case the gap does not arise from division by zero. Figure 4.7a shows that this gap may not be visible in the slope field; however, there is a visual clue to this particular pitfall and that is the sharp change in direction of the tangent lines near $x = 2$. This suggestion of a corner should provide a warning to look more carefully at the derivative equation.

The preceding examples suggest that the key to handling these confusing situations is to determine the domain of the derivative from the differential equation. There are three ways in which a function may fail to have a derivative:

- The derivative does not exist at a vertical tangent;
- The derivative does not exist at a corner;
- The derivative does not exist at any discontinuity.

The problem set that follows gives you practice finding solutions to initial value problems on slope fields, where the derivative fails to exist in one or more of the aforementioned ways.

Problems

www.peoplescollegeprep.com/slopefields.html

1. On a copy of the slope field in Figure 4.8, sketch the two solutions to the initial value problem $\sqrt[3]{x}\frac{dy}{dx} = 1$ passing through (2, 2) and (−3, 0).

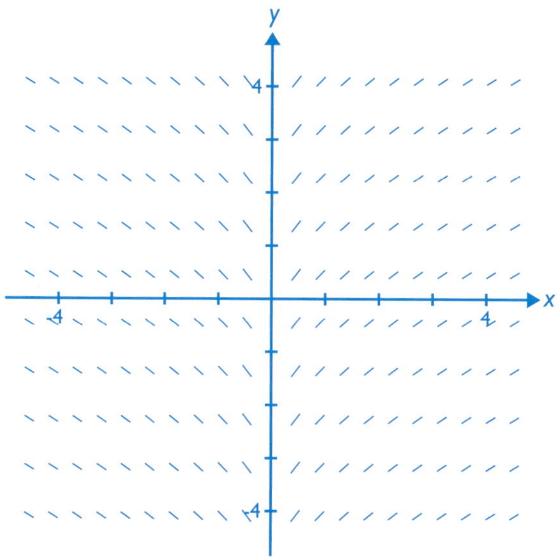

Figure 4.8

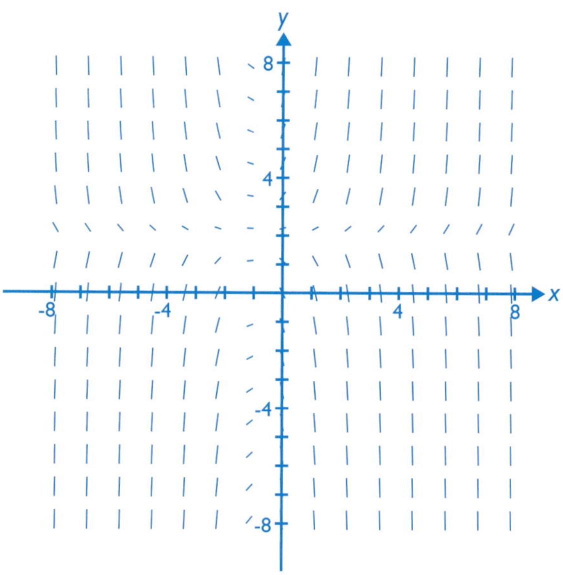

Figure 4.9

2. On a copy of the slope field in Figure 4.9, sketch the solutions to the initial value problem $\frac{dy}{dx} = (x + 1)(y - 2)$ passing through (0, 0), (2, 0), and (0, 4).

3. On a copy of the slope field in Figure 4.10, sketch the solutions to the initial value problem $\frac{dy}{dx} = \frac{e^{-x}}{y}$ passing through (0, −2) and (0, −4).

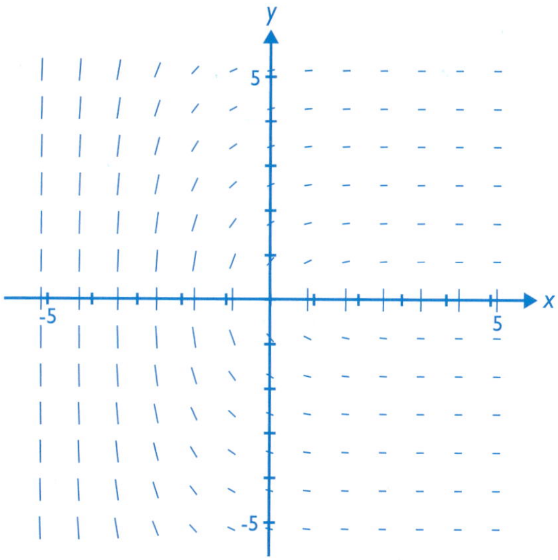

Figure 4.10

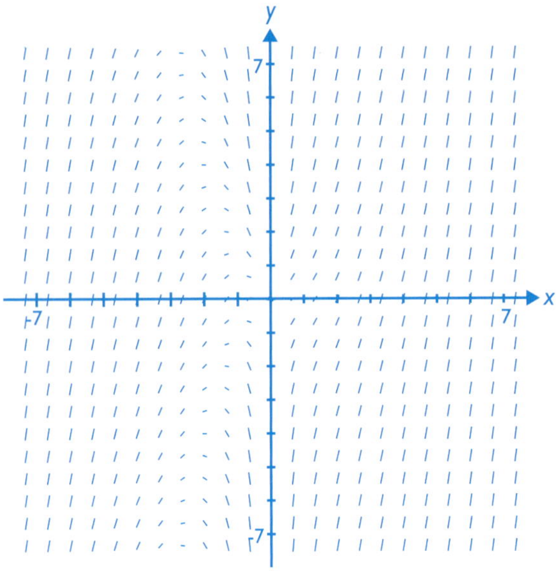

Figure 4.11

4. On a copy of the slope field in Figure 4.11, sketch the solutions to the initial value problem $\frac{dy}{dx} = |x| + \frac{|y|}{x}$ passing through $(-2, -2)$ and $(3, 3)$.

5. On a copy of the slope field in Figure 4.12, sketch the solutions to the initial value problem $\frac{dy}{dx} = \frac{1}{e^x + y}$ passing through $(0, 3)$ and $(0, -2)$.

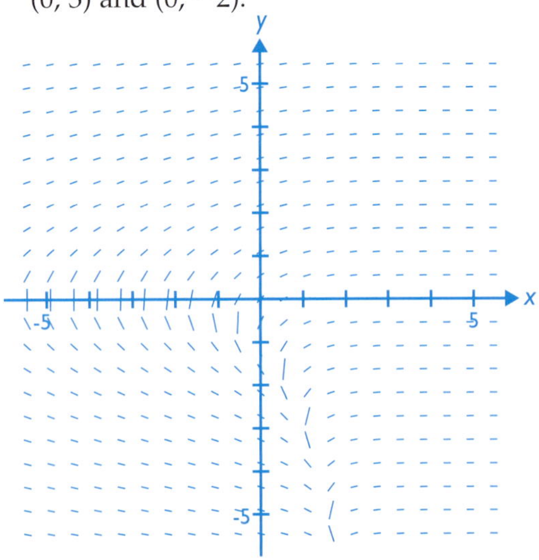

Figure 4.12

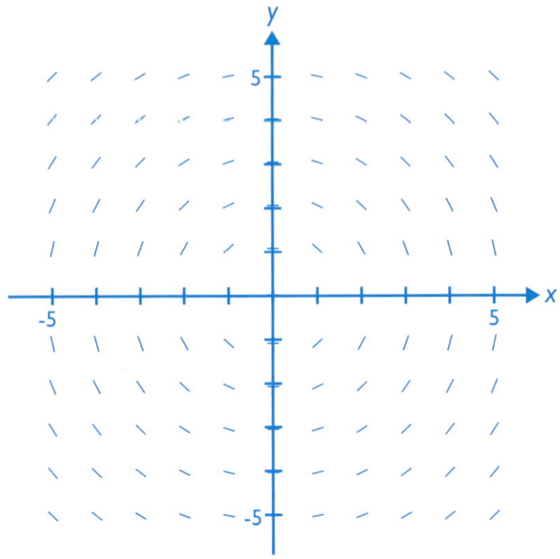

Figure 4.13

6. a. On a copy of the slope field in Figure 4.13, sketch the solution to the initial value problem $\frac{dy}{dx} = -\frac{x}{y}$ that passes through $(2, 2)$.

b. What is the equation of the solution curve you drew?

7. On a copy of the slope field in Figure 4.14, sketch the solution to the initial value problem $\frac{dy}{dx} = y + \frac{1}{x}$ that passes through $(4, -4)$.

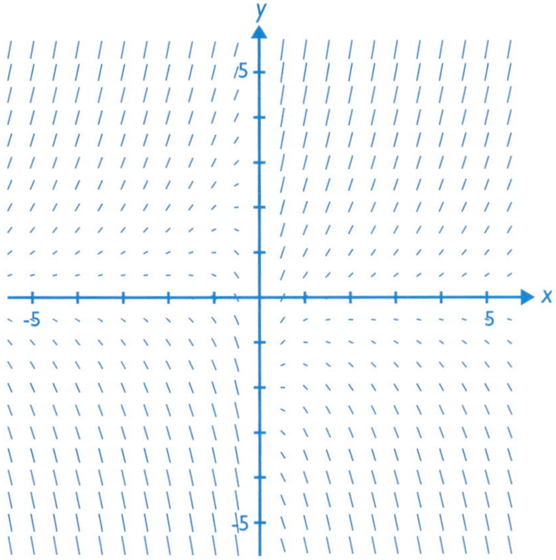

Figure 4.14

8. On a copy of the slope field in Figure 4.15, sketch the solutions to the initial value problem $\frac{dy}{dx} = \sin(x + y)$ that pass through $(0, 0)$ and $(0, -4)$.

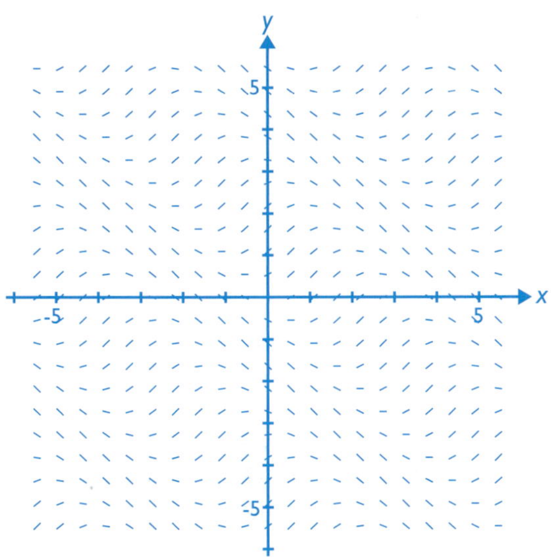

Figure 4.15

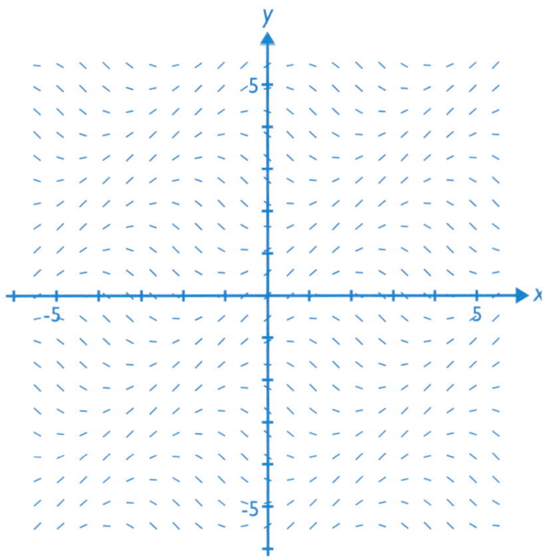

Figure 4.16

9. On a copy of the slope field in Figure 4.16, sketch the solution to the initial value problem $\frac{dy}{dx} = \cos(x - y)$ that passes through $(4, 0)$.

10. On a copy of the slope field in Figure 4.17, sketch the solutions to the initial value problem $\frac{dy}{dx} = (y - 2)(y - 4)(y - 1)$ that pass through (0, 2) and (0, 3).

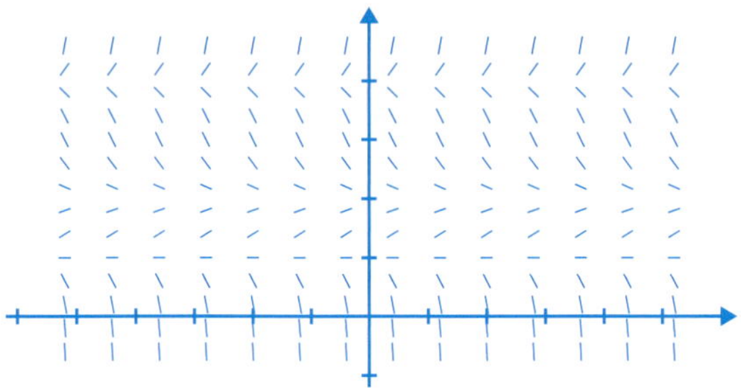

Figure 4.17

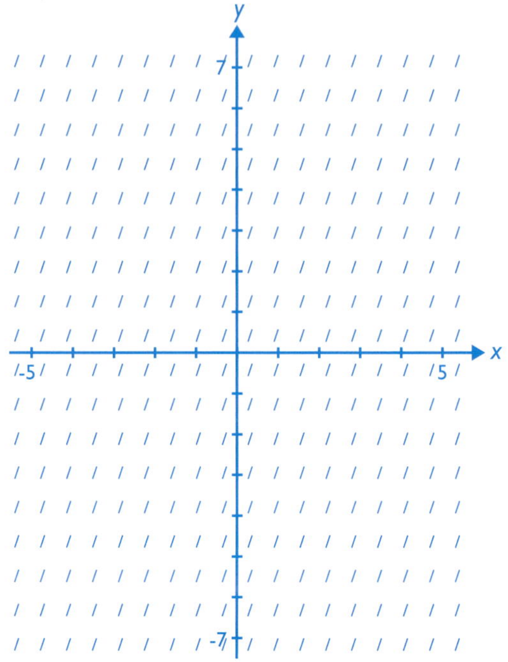

Figure 4.18

11. **a.** On a copy of the slope field in Figure 4.18, sketch the solutions to the initial value problem $\frac{dy}{dx} = \frac{3x + 9}{x + 3}$ that pass through (0, 4) and (−4, 4). (**Caution:** A computer-drawn slope field is not always perfectly accurate.)

b. Solve this differential equation algebraically. What is the domain and range of these particular solutions?

12. a. On a copy of the slope field in Figure 4.19, sketch the solution to the initial value problem $\frac{dy}{dx} = \frac{x(x-3)(x+2)}{x^2 - x - 6}$ passing through $(-1, 3)$. (**Caution**: A computer-drawn slope field is not always perfectly accurate.)

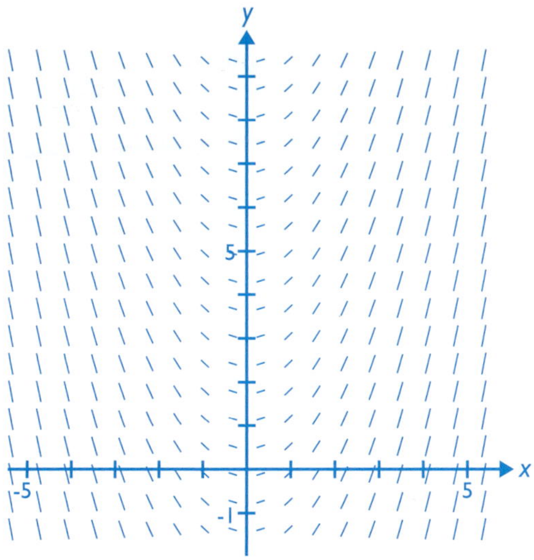

Figure 4.19

b. Solve this differential equation algebraically. What is the domain and the range of this particular solution?

CHAPTER 5

Stable and Unstable Equilibrium Solutions

It is time to examine more closely the boundary lines that have been noted in previous chapters. We pointed out, for instance, that the solution $y = -1 - x$ to $y' = x + y$ acted as a boundary line separating solutions into those above and those below this line. In another situation, we saw that both $y = -1$ and $y = 2$ were horizontal boundaries of solutions to $y' = (y - 2)(y + 1)$. There are two useful interpretations for such boundary lines.

First, as in each of the two examples just mentioned, the boundary lines are *asymptotes* of the solution curves; that is, they represent long-range behavior of the solutions. (Such is not always the case, as Example 1 below will show, but it is true in many cases.) When boundary lines are asymptotes, they take on special importance in applications where the long-range behavior is of equal or greater interest than specific values of the solution. A practical case in point is the analysis of the aftermath of an oil spill or other ecological disaster, where the long-range behavior is of critical importance. Will the body of water cleanse itself and return to its pre-spill condition or will it stabilize at some new level of contamination? This question is much more important than the particular amount of pollution that remains on any particular day.

Secondly, if the initial condition lies on one of these boundary lines, then the boundary line is itself a *particular solution*. In the special case in which this line is horizontal, the derivative is zero and the solution is constant. In this chapter we will focus on those differential equations that have such a horizontal boundary line. First, however, we will revisit a differential equation with a horizontal boundary line that is not an asymptote to its solution curves.

Example 1 Show that the x-axis is a boundary line to the solutions to the differential equation $\frac{dy}{dx} = \frac{1}{y}$.

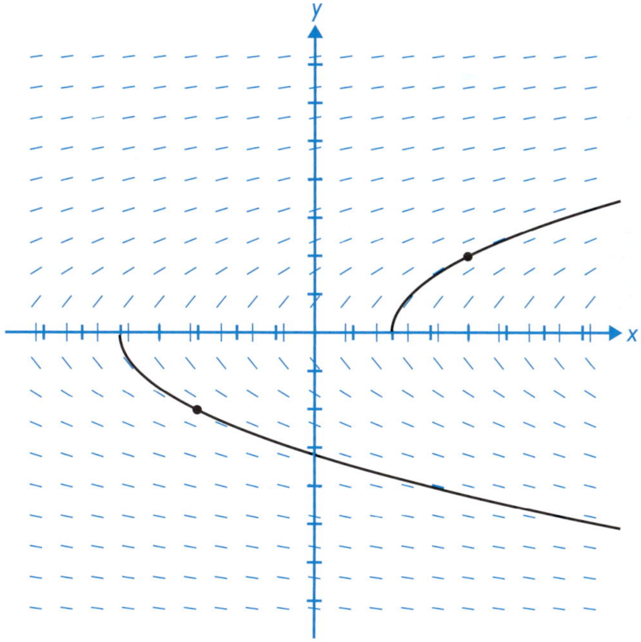

Figure 5.1

Solution Figure 5.1 shows a slope field and two particular solutions for this differential equation. Note that the derivative does not exist when $y = 0$. As solutions approach the x-axis, the derivative approaches infinity and the tangent lines approach the vertical. As we have seen, we cannot extend solutions across such a horizontal line, so the x-axis separates the solutions in the range $y > 0$ from those where $y < 0$. Note that the x-axis is not an asymptote to any solution.

Now let us move on to a differential equation with a boundary line that is itself a particular solution.

Example 2 Use a slope field to find a solution to the IVP $\frac{dy}{dx} = y - 2$ if $y = 2$ when $x = 6$.

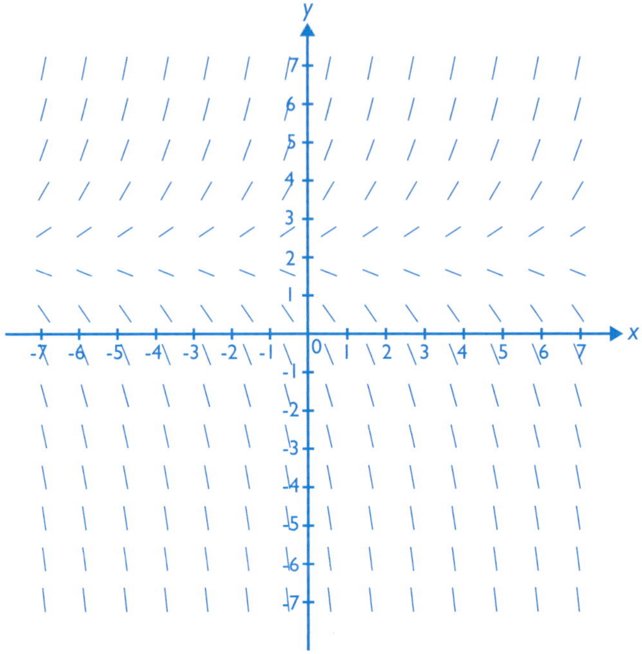

Figure 5.2

Solution Figure 5.2 shows a slope field for this differential equation, and it appears that the particular solution passing through (6, 2) is the line $y = 2$. We can confirm this answer analytically by showing that $y = 2$ is a solution to this differential equation, but it is instructive to think about the problem in another way. Imagine standing on (6, 2) and following the direction of the tangent line there. You will be led to another nearby point, which also lies on $y = 2$. Because the slope at that point is zero, if you continue moving in the direction of the tangent line you must remain on $y = 2$. We might call the latter form of reasoning *local thinking* because we restrict our attention to points near (6, 2). By contrast the algebraic solution, which shows that $y = 2$ satisfies this IVP, is *global* thinking because it deals with the whole solution all at once. Both types of thinking have their place in mathematics.

Example 2 illustrates an **equilibrium solution** to a differential equation. An equilibrium solution is constant for all values of the independent variable. Because a constant function has zero slope, we can find equilibrium solutions by setting the derivative equal to zero or by looking for rows of horizontal tangent segments in a slope field.

Example 3 Find the equilibrium solutions to $\frac{dy}{dx} = y^2 + 3y - 18$.

Solution Set $y^2 + 3y - 18 = 0$ and solve to get $y = -6$ and $y = 3$ as the equilibrium solutions.

As we have seen, one of the properties of equilibrium solutions is that they represent boundaries that separate the plane into different regions with different solution types in each region. A second property of these equilibria is that they are the only constant functions that solve the differential equation. In the previous example, if $y = k$ is a constant function that satisfies the differential equation, then $\frac{dy}{dx} = k^2 + 3k - 18$ and this derivative must be identically zero, because the derivative of a constant is zero. Therefore, $k^2 + 3k - 18 = 0$ and $k = -6$ or $k = 3$.

A third property of these equilibria is that in many situations they are horizontal asymptotes of the solution curves as x approaches ∞. When this occurs, solutions get closer and closer to the equilibrium solution. The opposite behavior is possible as well and some equilibria are characterized by solution curves moving farther and farther away from the equilibrium as x increases. In the former case the equilibrium is called **stable**. In the latter case it is called **unstable**.

This vocabulary is borrowed from Physics. Imagine, for example, a book lying on a table. The book is said to be in equilibrium because it is at rest. If we lift one corner of the book and release it the book will fall back to the table. Since it moves towards its equilibrium position, that state of rest is called stable. Imagine instead that that book is balanced on its edge. This too is an equilibrium position since the book is not moving. However, it is in an unstable equilibrium since the book will fall over and move away from equilibrium if we push it slightly off-balance.

We will extend this language to the solution curves as well and call solutions that move towards equilibrium as x increases **stable solutions,** while those that move away from equilibrium will be called **unstable solutions**.

Example 4 Use the slope field in Figure 5.3 to identify the stable and unstable equilibrium solutions. What initial conditions will produce solutions that move toward equilibrium? What initial conditions will produce solutions that move away from equilibrium? Are there initial conditions that do neither?

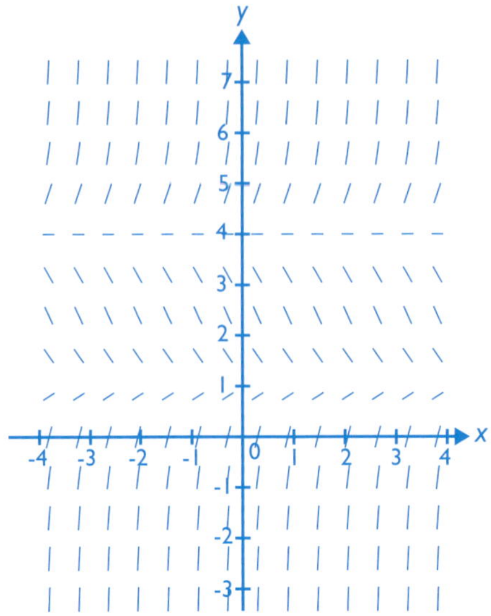

Figure 5.3

Solution The slope field suggests that $y = 4$ and $y = 1$ are equilibrium solutions. The line $y = 4$ is thus an unstable equilibrium because solution curves, both above and below, move away from it as $x \to +\infty$. The line $y = 1$, on the other hand, is a stable equilibrium because solution curves move toward it as $x \to +\infty$. Initial conditions with $y > 4$ will always produce a solution that veers away from the unstable solution $y = 4$, and initial values with $y < 1$ will always produce solutions that move toward $y = 1$. Initial conditions with $1 < y < 4$ produce curves that do both—they move away from the unstable equilibrium at $y = 4$ and toward the stable equilibrium at $y = 1$.

Any initial condition with $y = 4$ gives the constant solution $y = 4$, and any initial condition with $y = 1$ gives the constant solution $y = 1$. These are the two equilibrium solutions.

Summary

- To find equilibrium solutions in a slope field, look for rows of horizontal tangents.
- To find equilibrium solutions algebraically, solve for $y' = 0$.

Problems

www.peoplescollegeprep.com/slopefields.html

1. Figure 5.4 shows a slope field for the differential equation $\frac{dy}{dx} = y^2 - 8$. There appear to be two equilibrium solutions. Find these algebraically and decide whether each is stable or unstable.

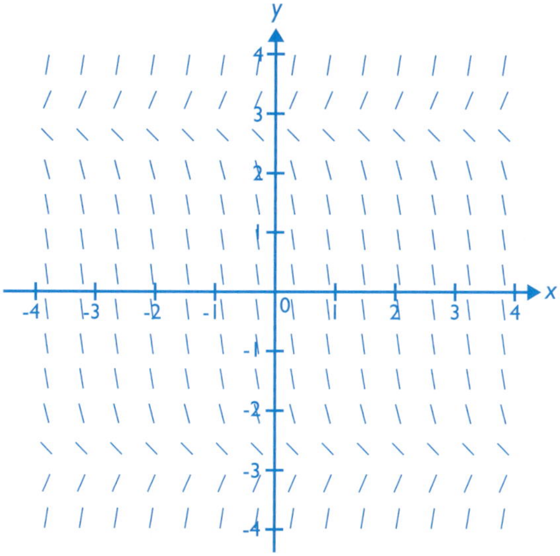

Figure 5.4

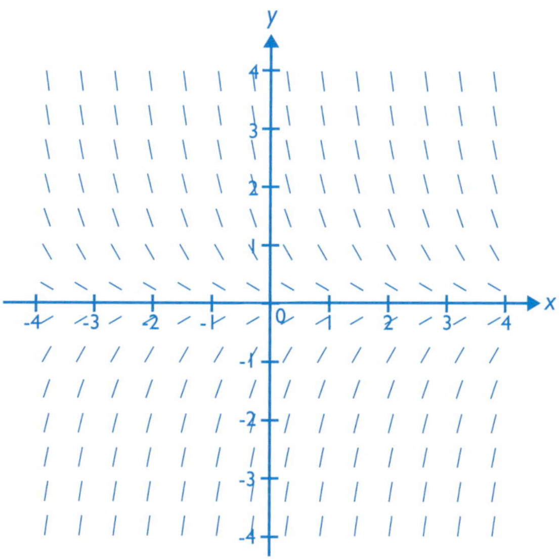

Figure 5.5

2. Find the equilibrium solutions to $\frac{dy}{dx} = -2y$ and decide whether each is stable or unstable. Figure 5.5 shows the slope field for this differential equation.

3. Figure 5.6 shows the slope field for some differential equation. Sketch the solution passing through (0, 5) and the solution through (2, 7).

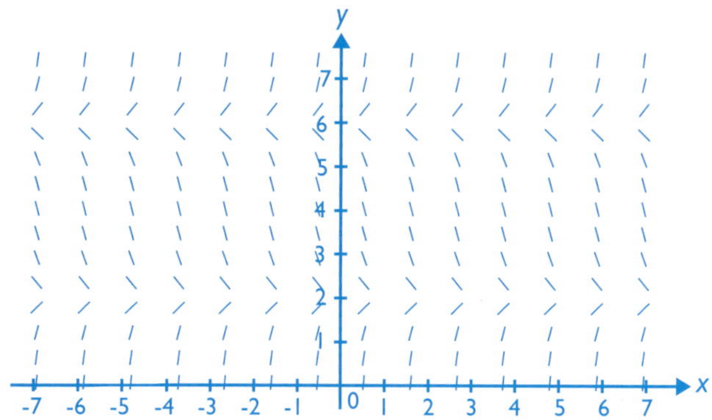

Figure 5.6

4. Figure 5.7 shows the slope field for some differential equation.

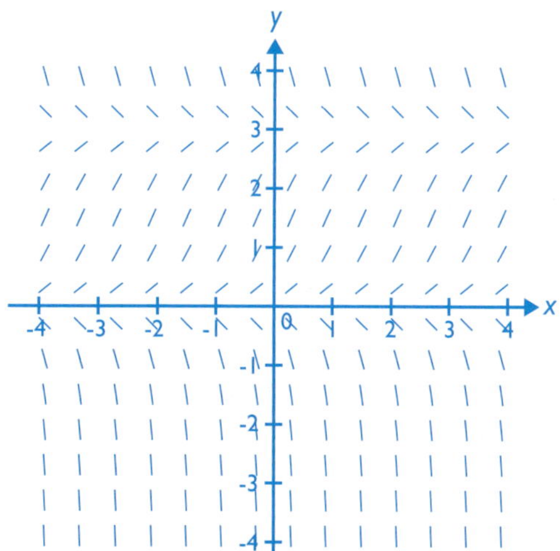

Figure 5.7

   a. Sketch the solution passing through $(-2, 2)$ and the solution through $(-2, 3)$.

   b. Approximate the equilibrium solution(s) and decide if each is stable or unstable.

5. $f(x) = 46 - 25e^{-0.025x}$ is a solution to some initial value problems with initial condition $f(0) = 21$. Does the differential equation have a stable or an unstable equilibrium? Explain.

6. $y = \dfrac{\sqrt{3x}}{\sqrt{x-4}}$ is a solution to some initial value problem with initial condition $y(0) = 0$. Does the differential equation have a stable or an unstable equilibrium? Explain.

7. a. $f(x) = (x - 4)(x + 1)$ is a solution to some initial value problem with initial condition $f(2) = -6$. Does the differential equation have a stable or an unstable equilibrium? Explain.

8. Find the equilibrium solutions, if any, to the differential equation

$$\frac{dy}{dx} = 3y(y - 2)(y - 18).$$

9. Find the equilibrium solutions, if any, to the differential equation

$$\frac{dy}{dx} = 3x(x - 2)(x - 18).$$

10. Find the equilibrium solutions, if any, to the differential equation

$$\frac{dy}{dx} = (x - 3)(y + 8).$$

11. a. Find the equilibrium solutions, if any, to the differential equation

$$\frac{dy}{dx} = -y^2 + 9y - 14.$$

b. Differentiate this equation implicitly and substitute to get y'' as a function of y alone. Use the second derivative to identify the location of the inflection points.

c. What range of initial values will produce a solution with such an inflection point?

12. a. Find the two equilibrium solutions to the differential equation $\frac{dy}{dx} = y^3 - 6y^2$.

b. Examine a slope field for this equation in a window that shows both equilibrium solutions. One of them is clearly either stable or unstable. Which solution can be easily identified? Is it stable or unstable?

c. The other solution appears to be both stable and unstable. Find the range of solutions for which this second equilibrium is stable and the range for which it is unstable.

13. Figure 5.8 is a slope field for the differential equation

$$\frac{dy}{dx} = (x + y + 4)(x + y + 2).$$

It seems to have a boundary line that separates the solutions into two families. Show that $y = -x - 3$ is the boundary line.

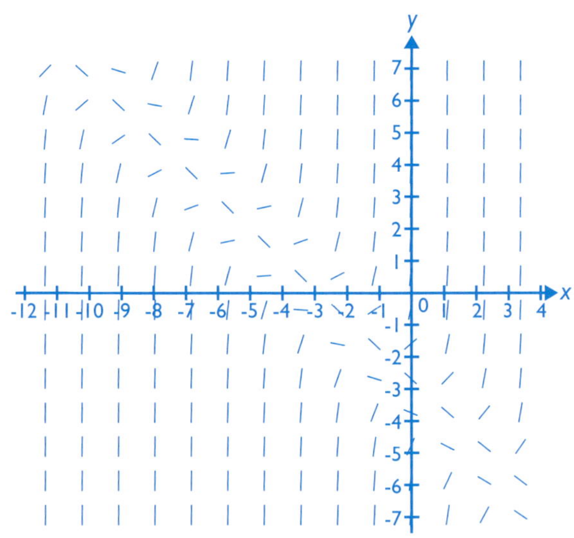

Figure 5.8

14. a. On a copy of the slope field in Figure 5.9, sketch the solutions to the initial value problem $\frac{dy}{dx} = x(y - 2)$ passing through (0, 4), (2, 4), and (4, 4).

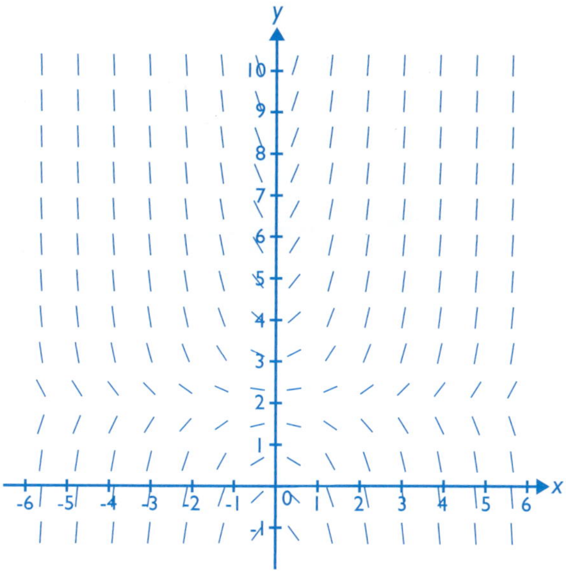

Figure 5.9

b. The only equilibrium solution is $y = 2$. Is this a stable or an unstable equilibrium?

c. The particular solution curves in **14a** are U-shaped. Does this contradict our earlier observation that such boundaries are asymptotes? Explain.

CHAPTER 6

When the Derivative Is a Function of One Variable

Slope fields can be a wonderful aid in trying to understand the properties of the solution to a differential equation, but they are not the only tool we can bring to bear on the problem. In this chapter we will look at another type of graph that can tell us much about the solution to a differential equation when the derivative is a function of a single variable.

The differential equation in the first example should be very familiar. You probably solved many similar problems when you first began studying calculus.

Example 1 Use the graph of the derivative $\frac{dy}{dx} = x^2 - 4$ in Figure 6.1 to describe a possible graph for the function y.

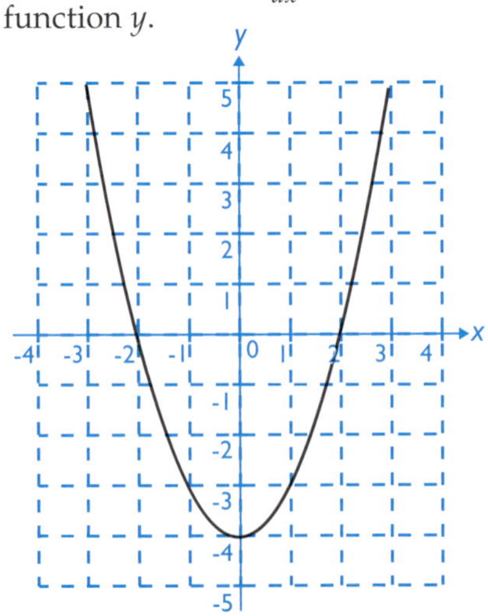

Figure 6.1

Solution

Our aim is to use the features of the graph of the derivative to find a solution to a differential equation. Of course, this particular differential equation would be very simple to solve algebraically, but it is useful to recall how to solve it graphically.

Let us first look at Figure 6.2, a **sign diagram** for the derivative, which is positive for $|x| > 2$, negative for $|x| < 2$, and zero for $x = \pm 2$. We know therefore that the solution will increase for $|x| > 2$, decrease for $|x| < 2$, and have a local maximum at $x = -2$ and a local minimum at $x = 2$.

Figure 6.2

We can get more information from the second derivative, $\frac{d^2y}{dx^2} = 2x$, a function that is negative for $x < 0$ and positive for $x > 0$. This tells us that the solution-curves are concave down for negative x and concave up for positive x. Since the concavity changes at $x = 0$, that will be a point of inflection of the solution. Notice how well these properties of a solution describe the family of cubics that are the antiderivatives of x^2.

Now a slight change to the differential equation in Example 1 will result in a very different analysis.

Example 2 Analyze the solutions to the differential equation $\frac{dy}{dx} = y^2 - 4$.

Solution

This is a separable equation and it is possible to find exact solutions. (In fact, if you have studied **logistic growth**, you may already know how to solve this equation algebraically.) However, this derivative is so similar to the one in Example 1 that it is useful to compare a similar graphical approach to its solution. Figure 6.3 shows a graph of this derivative. Note that it is identical to Figure 6.1 except that the horizontal axis is y. This simple difference makes all the difference. A sign diagram for this function is shown in Figure 6.4, identical to 6.2 except that the independent variable is now y.

It is easy to misread such a graph since we are so used to x being on the horizontal axis, so we must take special care to interpret the figure correctly. When $y = \pm 2$ the derivative is zero, but instead of turning points these are the equilibrium solutions of the differential equation. That is, these horizontal lines separate the solutions into three families. The solutions where $|y| > 2$, that is, those below $y = -2$ and those above $y = 2$, are increasing functions since the derivative is positive in those regions. The solutions where $|y| < 2$, that is, those between the equilibria, are decreasing since the derivative is negative there.

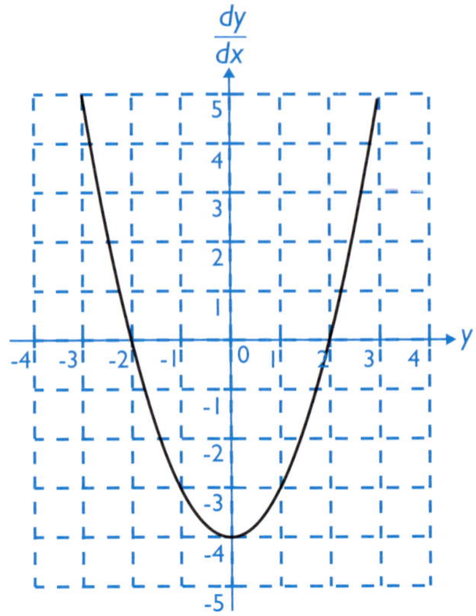

Figure 6.3

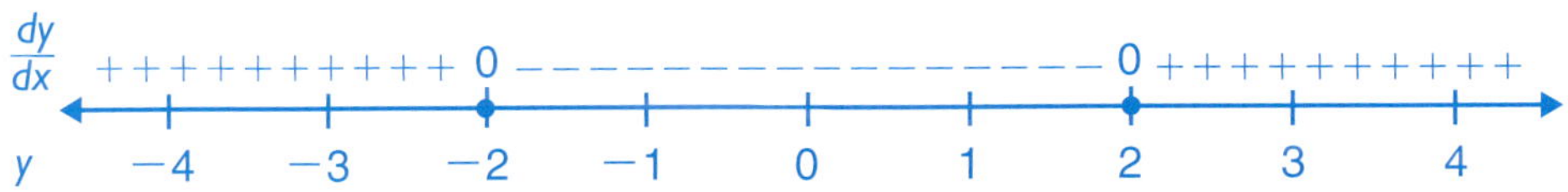

Figure 6.4

We can get more information by differentiating the equation implicitly to get $\frac{d^2y}{dx^2} = 2y\frac{dy}{dx}$. Substitute the equation for the derivative to get $\frac{d^2y}{dx^2} = 2y(y^2 - 4)$. This derivative changes sign at $y = -2, 0$, and $+2$, as shown in Figure 6.5, a sign diagram for this cubic.

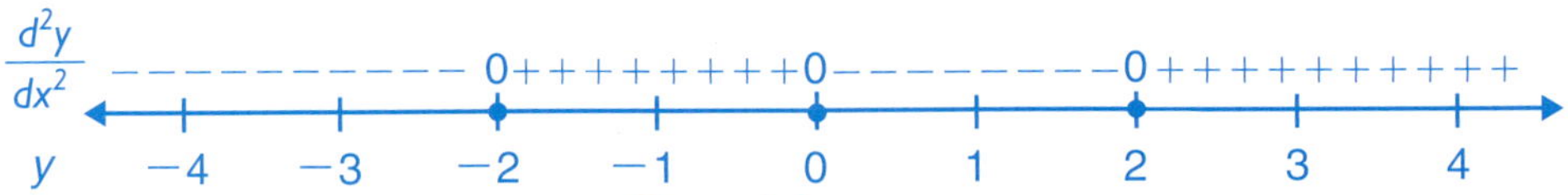

Figure 6.5

In Example 1, this analysis applied to a single function. Here though, the sign diagram describes the concavity of all three families of solutions. The solutions below $y = -2$ are increasing and concave down, the solutions above $y = 2$ are increasing and concave up, and the solutions between the equilibria are decreasing and concave down below $y = 0$ and then decreasing and concave up below $y = 0$. Figure 6.6 illustrates these families with solutions passing through (1, 3), (1, 1), and (1, −3). Note that none of the solutions have maxima or minima and that the solution through (1, 1) has a point of inflection on the x-axis. In general, every solution in the family of curves between the equilibria has a point of inflection on the x-axis as well.

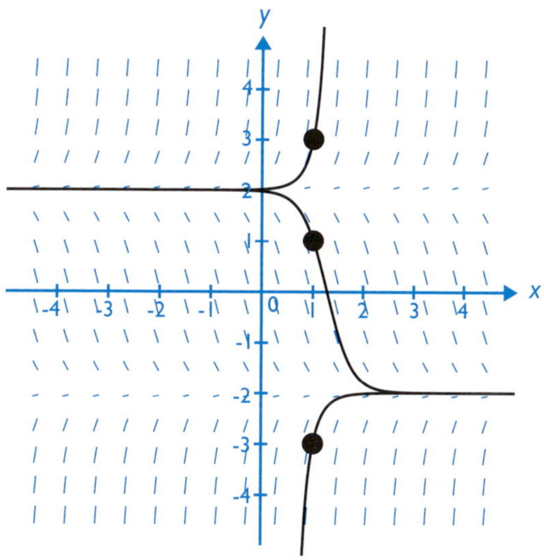

Figure 6.6

A third example will clarify this type of analysis. Study these examples carefully so that you can try your hand at similar problems at the end of the chapter.

Example 3 Analyze the types of solutions possible for $\frac{dy}{dx} = \cos(y)$.

Solution Again we graph the derivative against y, getting the cosine curve shown in Figure 6.7. The range of the cosine tells us that the slope of the solution curve is never greater than $+1$ nor less than -1. The zero values of the derivative at $y = \frac{\pi}{2} + n\pi$ are, again, equilibrium solutions. That is, the solutions fall within infinitely many horizontal bands of width π.

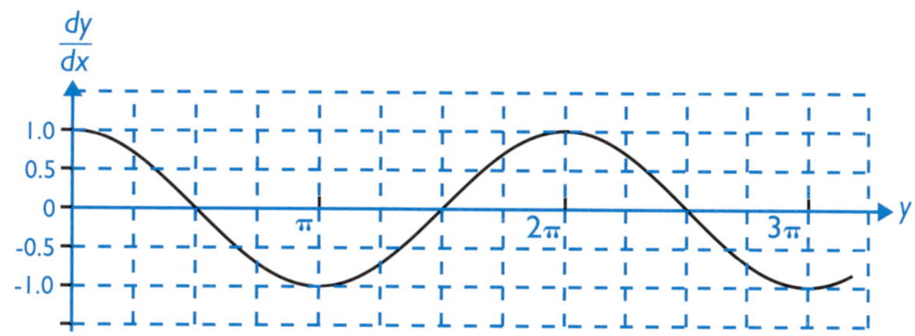

Figure 6.7

The signs of the cosine tell us that the solutions increase in quadrants I and IV and decrease in quadrants II and III. The maxima and minima of the cosine occur at $y = n\pi$, where the slopes of the cosine change sign. Therefore, the solution curves will have points of inflection that lie on the lines $y = n\pi$.

Figure 6.8 shows a slope field for this differential equation, with the equilibrium solutions $y = -\frac{\pi}{2}, \frac{\pi}{2}$ and $\frac{3\pi}{2}$ drawn as well as the particular solutions through (0, 1), (2, 2) and (−3, −3). Note that in this graphic the solutions are seen between the now horizontal equilibrium solutions at $y = \frac{\pi}{2} + n\pi$, and the points of inflection are on the horizontal lines $y = n\pi$.

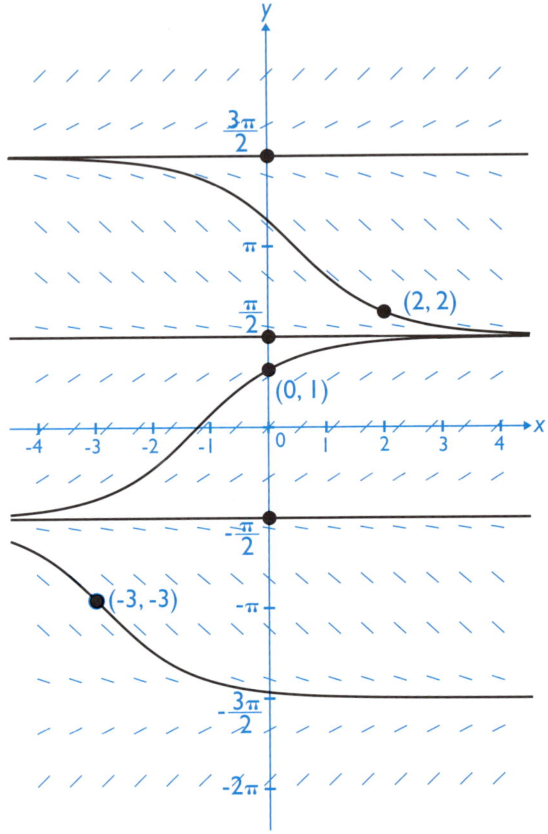

Figure 6.8

The next example illustrates how useful this kind of derivative graph can be in a problem applied to the spread of a disease. The situation is based upon this simple idea: healthy people get sick and sick people get well. Each sick person is a potential source of the disease, spreading it to healthy people; each healthy person is a potential victim of the disease, who may catch it. One model for this situation is called the **logistic model** and assumes that the rate at which a disease spreads (the rate at which people sicken) is proportional to both the number of healthy people and the number of sick people. Let $P(t)$ be the number of sick people in a population of size C at time t. We can model this situation algebraically by $\frac{dP}{dt} = kP(C - P)$.

Example 4 Use a graph of $\frac{dP}{dt}$ against P to analyze the types of solutions possible in this model of spread of disease. Assume $k > 0$.

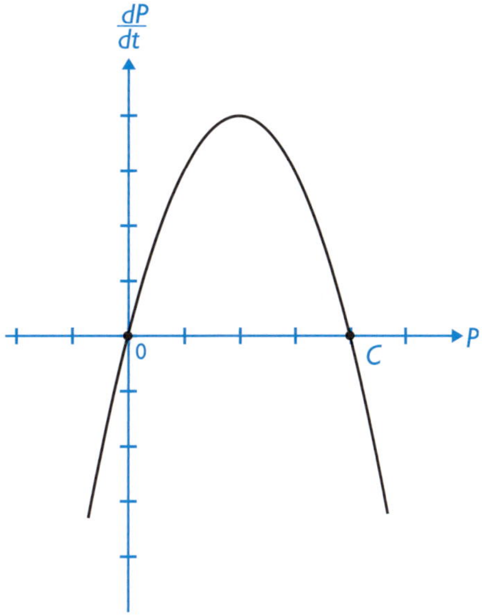

Figure 6.9

Solution Figure 6.9 shows a graph of $\frac{dP}{dt}$ against P. It is a parabola with intercepts at $P = 0$ and $P = C$, so these are the equilibrium solutions. They correspond to the uninteresting extremes where either everybody is sick ($P = C$), and so there is nobody left to catch the disease, or where nobody is sick ($P = 0$), and there is nobody left to spread the disease. In both cases $P' = 0$. Since it is not possible for $P < 0$ or $P > C$, the only solutions will be between the equilibria, where the derivative is positive. Therefore, the number of sick people will increase steadily from its initial value and level off approaching $P = C$.

Now differentiate implicitly to find the second derivative:
$\frac{d^2P}{dt^2} = k(C - 2P)\frac{dP}{dt}$. The second derivative is zero at the equilibria and when $P = \frac{C}{2}$. Only the last value is interesting, since it is within the range of values of P that we are considering. The change of sign at $P = \frac{C}{2}$ marks a point of inflection of the solution. This appears as a turning point in Figure 6.9, where the derivative $\frac{dP}{dt}$ is graphed. Therefore, the number of sick people will increase at an increasing rate until half the population is sick, at which point the disease will continue to spread but at an ever decreasing rate.

Remember: In general, if a differential equation is a function of y alone, the graph of the derivative against y is a nice, informative complement to a slopefield graph.

Problems

www.peoplescollegeprep.com/slopefields.html

In Problems 1–8 use a graph of the derivative to analyze the solutions to each differential equation. For each solution type find the intervals on which it increases and decreases, where it is concave up and concave down, and identify the locations of any points of inflection. You do not need to solve these equations exactly.

1. $\frac{dy}{dx} = \frac{y^2}{2}$

2. $\frac{dy}{dx} = 3y + 4$

3. $\frac{dy}{dx} = 3(y - 300)(y + 200)$

4. $\frac{dy}{dx} = 400y(1000 - y)$

5. $\frac{dy}{dx} = x(x - 3)(x + 5)$

(Note that this derivative is a function of x only.)

6. $\frac{dy}{dx} = e^{-y^2}$

7. $\frac{dy}{dx} = ye^y$

8. $\frac{dy}{dx} = 3\sin(\pi y)$

9. The **logistic model of growth** was first proposed by the Belgian mathematician Pierre Verhulst in 1845. He reasoned that a population can only grow while there is available food and shelter and that once those quantities diminish so will the population. Verhulst supposed that any particular environment could carry a maximum population C, which he called the **carrying capacity** of the environment. He assumed that the rate of growth of a population is proportional to both the population size, P, and also to the unused remaining capacity of the environment, $C - P$. That is, $\frac{dp}{dt} = kP(C - P)$. This is exactly the same differential equation we found in Example 4 when modeling the spread of disease.

a. Use the methods of this chapter to analyze the possible solutions for a population of deer living on an island that has a carrying capacity of 1200 deer. Suppose $k = 0.005$.

b. Can any meaning be attached to solutions where the initial value is greater than 1200?

10. The logistic differential equation can also be used to model the spread of a rumor in a population. It seems reasonable to think the rate of spread of the rumor will be proportional to the number of people who have heard the rumor (since they are the ones who can pass the rumor on) and also proportional to the number of people who have *not* heard the rumor.

a. Suppose Central High has 800 students and that 2 students start a rumor on day 0. Write a logistic differential equation with initial conditions to model this situation.

b. Since we do not know the value of k we cannot make a precise graph for this situation, but we can still make a qualitative graph, one that displays the information that we do know. Let $R(t)$ be the number of students who have heard the rumor by day t. Sketch a qualitative graph of $\frac{dR}{dt}$ against R showing the critical features of the curve.

11. If $\frac{dy}{dx} = y^2$ find a formula for $\frac{d^n y}{dx^n}$.

12. Figure 6.8 shows the graph of $\frac{dy}{dx}$ against y for some differential equation. Use this graph to sketch a possible solution to the differential equation passing through (0, 0).

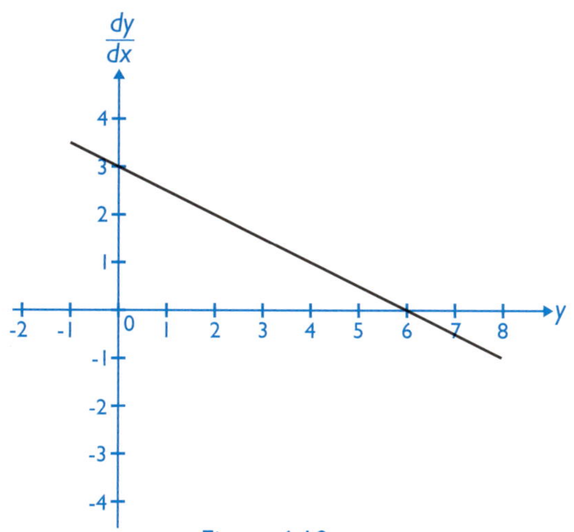

Figure 6.10

13 Figure 6.9 shows the graph of $\frac{dy}{dx}$ against y for some differential equation.

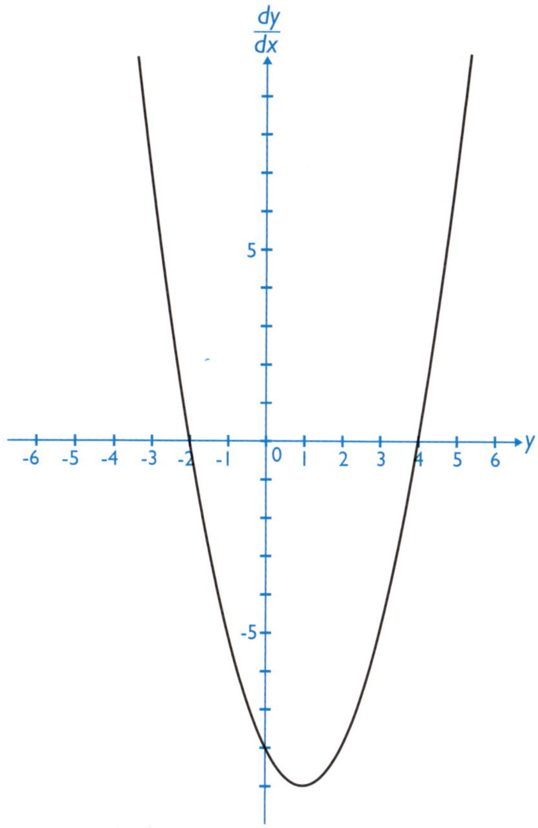

Figure 6.11

a. Use this graph to sketch a possible solution to the differential equation passing through (0, 0).

b. Use this graph to sketch a possible solution to the differential equation passing through (2, 6).

c. Use this graph to sketch a possible solution to the differential equation passing through $(-2, -4)$.

CHAPTER 7

Numerical Solutions: Euler's Method

Many differential equations can be solved exactly either by direct antidifferentiation or by other means. For example, $y' = x^3$ can be solved directly and $y' = xy$ can be solved by separation of variables. When an antiderivative is not known, we can still *approximate* values of the solution to differential equations that are functions of x, using the Second Fundamental Theorem of Calculus. However, many differential equations, even apparently simple ones like $\frac{dy}{dx} = x + y$, cannot be solved exactly without more advanced techniques. As we have seen, solutions to such equations can be visualized using slope fields. But while slope fields may give us a general idea of the types of solutions possible and some of their features, they do not let us compute specific values of the solutions with any accuracy.

Various methods have been created to approximate solutions in these cases. In this chapter, we will examine one method devised by the Swiss mathematician Leonhard Euler (1707-1783). **Euler's Method** is based on a tool you already know well, tangent line approximations.

Suppose, for example, we would like to know the value at $x = 2$ of the solution to the differential equation $y' = x + y$ passing through (1, 1). Because we cannot find an explicit formula that would allow us to compute $y(2)$, we settle for the next best thing, a method that will approximate it. We know that we can use the tangent line at (1, 1) to approximate $y(2)$ at least roughly, since (1, 1) is a solution. The derivative at (1, 1) is 2,

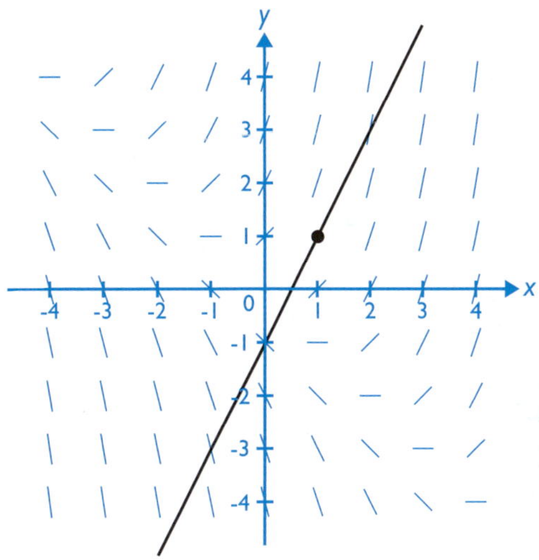

Figure 7.1

and so the equation of this tangent line is $y = 2x - 1$. Figure 7.1 shows a slope field for the differential equation and the tangent line at (1, 1). Note that this line is not a solution but an exaggeration of one of the tangent segments in the slope field. When $x = 2$, the y-value on this tangent is 3, and so we can make the approximation $y(2) \approx 3$.

This value is not likely to be a very good approximation, however. Tangent lines are good approximators when we stay close to the point of tangency. Using the same tangent line to approximate $y(1.5)$, for example, probably yields a better approximation than our approximation for $y(2)$. That is, the tangent line at (1, 1) will better approximate the solution at $x = 1.5$ than the solution at $x = 2$. Euler's clever idea was to think of using this intermediate approximation, $y(1.5) \approx 2(1.5) - 1 = 2$, as a stepping stone on the way to $x = 2$.

Of course, (1.5, 2) is not a point on the solution, *but it is closer to the graph* than (2,3). Now Euler repeated the process using a tangent drawn at (1.5, 2) and using the derivative to compute the slope at this new point.

Following Euler, we repeat the process and use the derivative to approximate the slope at (1.5, 2) to get $y'(1.5) \approx 3.5$. The equation of a tangent line at (1.5, 2) is $y = 3.5(x - 1.5) + 2$. Now we use this *second* tangent line to approximate: $y(2) \approx 3.5(2-1.5) + 2 = 3.75$. Figure 7.2 shows the solution through (1, 1) as well as both our approximating solutions. Notice that (2, 3.75) is closer to the solution curve than (2, 3).

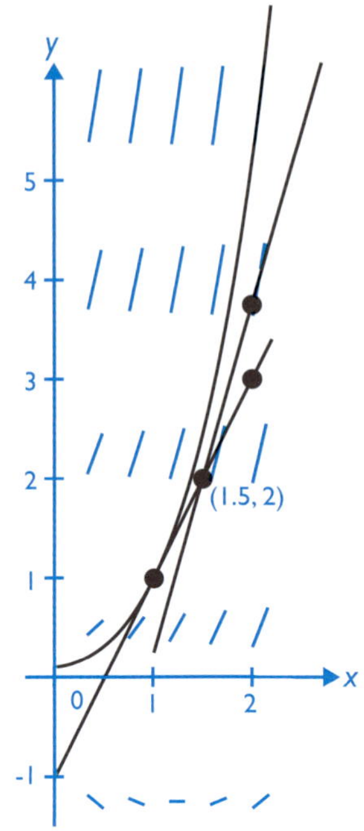

Figure 7.2

Let us review Euler's logic for a moment. The tangent line approximation at (1, 1) predicts a value of $y(2) \approx 3$. By using two tangent lines instead of one, we get the approximation $y(2) \approx 3.75$. Of course, if two tangent lines are better, why not use three or four or ten? Because each tangent line requires additional computation, there is a trade-off between the accuracy we gain and the drudgery of the computation.

Fortunately, the method is not difficult to program. You will find a calculator program for Euler's Method that will be useful. Appendix C shows how to implement Euler's Method using an Excel spreadsheet. In the meantime, the following example will illustrate the general method.

Example 1 Approximate the value of the solution to $\frac{dy}{dx} = \frac{1}{x} + \frac{1}{y}$ at $x = 2.5$. Start at (1, 1) and use $\Delta x = 0.5$.

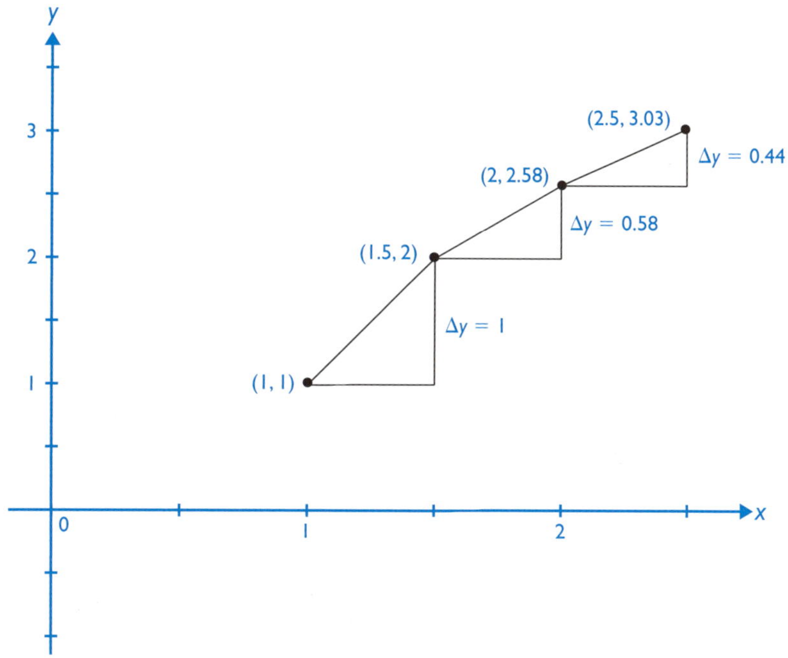

Figure 7.3

Solution Figure 7.3 illustrates the first three steps of Euler's Method. We begin at (1, 1) and draw a tangent line to (1.5, 2). From (1.5, 2) we draw a tangent line to (2, 2.58), and from there a third tangent line takes us to (2.5, 3.03). So $y(2.5) \approx 3.03$.

Here are details of the procedure. Begin at (1, 1) and compute $y' = 2$. Because $\frac{\Delta y}{\Delta x} = \frac{\Delta y}{0.5}$, $\Delta y = 1$. This computation gives coordinates of the next point, (1.5, 2).

Repeat the process for this point: $y'(1.5,\ 2) \approx 1.167$, so that the next $\Delta y = 0.583$. This gives the third point (2, 2.583) with $y'(2, 2.583) \approx 0.887$, so that $\Delta y = 0.444$ and the next point is (2.5, 3.027).

Because the computations do get messy, it is best to organize them in a table like the one displayed in the next example.

Example 2 Approximate the value of $y(3)$ on the solution to the IVP $\frac{dy}{dx} = 2x - y + 1$, where $y(1) = -4$. Use Euler's Method with 8 steps.

Solution Table 7.1 shows successive computations. We want to move from $x = 1$ to $x = 3$ in 8 steps, so our step size is $\Delta x = 0.25$. The first column lists successive x-values beginning with $x = 1$, so this column has the values x_0, $x_0 + \Delta x$, $x_0 + 2\Delta x$, $x_0 + 3\Delta x + \ldots$.

x	y	$\Delta y = y'(x)\Delta x$
1	−4	1.75
1.25	−2.25	1.4375
1.5	−0.8125	1.203125
1.75	0.390625	1.02734375
2.0	1.41796875	0.895507813
2.25	2.31347656	0.796630859
2.5	3.11010742	0.722473145
2.75	3.83258057	0.666854858
3.0	4.4995	

Table 7.1

The second column lists successive y values, and the last column gives the approximate change in y based upon a tangent line approximation at the current point. Because the slope of the tangent line is $y'(x) \approx \frac{\Delta y}{\Delta x}$, the change in y is $\Delta y \approx y'(x)\Delta x$. Notice that each y-value, after the initial value, is the sum of the previous y-value and the previous Δy. One computes one row at a time.

Problems

www.peoplescollegeprep.com/slopefields.html

1. a. Use Euler's Method with $\Delta x = 0.5$ to approximate $f(3.5)$ where $f(x)$ satisfies the initial value problem $f'(x) = x$ and $f(2) = 2$.

   b. Solve the IVP and use your solution to find the exact value of $f(3.5)$.

2. a. Using Euler's Method with five steps and $\Delta x = 0.2$, approximate $y(2)$ of the solution to the differential equation $y' = x - y$ that passes through $(1, 0)$

   b. Use Euler's Method with five steps to approximate $y(2)$ for the same equation if instead the solution passes through $(3, 3)$.

3. a. Use Euler's Method with eight steps and $\Delta x = 0.2$ to approximate $y(2.6)$ if $\frac{dy}{dx} = \sqrt{xy}$ and the solution passes through $(1, 1)$.

   b. Solve this same initial value problem exactly and compute the exact value of the solution at $x = 2.6$.

4. a. Use Euler's Method with 10 steps to approximate $y(4)$ on the solution to the differential equation $y' = (y - 2)(y - 8)$ passing through $(3, 2)$.

   b. Explain why Euler's Method gives the exact value in this case.

5. a. Sketch the solution to the initial value problem $\frac{dy}{dx} = 1 - x - y$, starting at $(-2, 3)$, using a copy of the slope field in Fig. 7.4. Then use your sketch to approximate $y(-1)$.

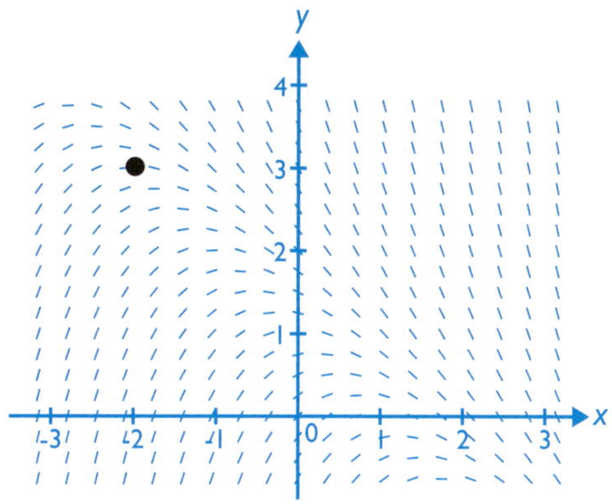

Figure 7.4

   b. Approximate $y(-1)$ using Euler's Method with two steps. Is your approximation an overestimate or an underestimate?

   c. On a copy of the slope field in Figure 7.5, repeat your sketch from **5a** and draw the two tangent lines and the three points used in Euler's Method.

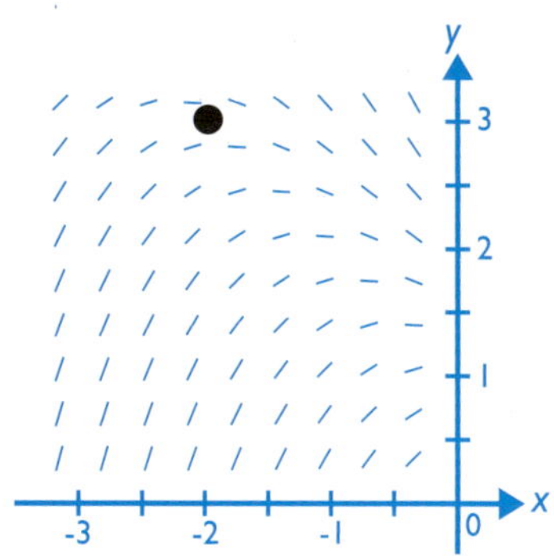

Figure 7.5

6. a. Figure 7.6 shows a graph of $y = \log_{10}(x)$. On a copy of the figure, draw the tangent line at $x = 10$ and use it to approximate $\log_{10}(2)$.

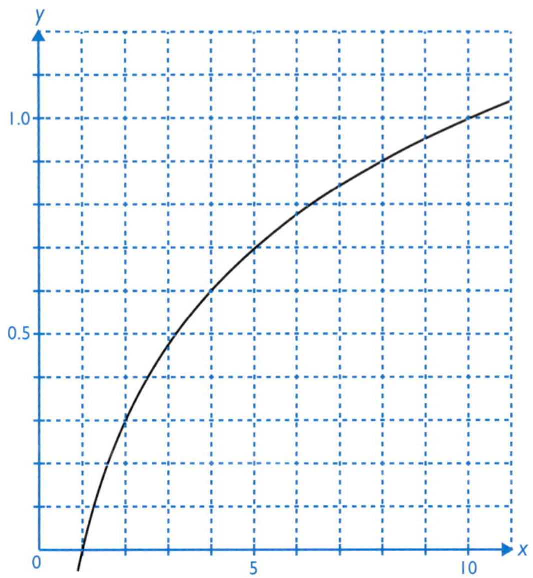

Figure 7.6

b. Use the line you drew for **6a** to approximate $\log_{10}(6)$ and then draw a line from the new point with the correct slope of $y = \log_{10}(x)$ at $x = 6$. Use this line to get a second approximation for $\log_{10}(2)$. Graphically, which approximation is closer to the correct value?

c. Use your calculator to find the error in each approximation.

7. a. Figure 7.7 shows a graph of $g(x) = 8x^3 - 8x$. On a copy of the figure draw the tangent line at $x = -0.5$ and use it to approximate $g(0.5)$.

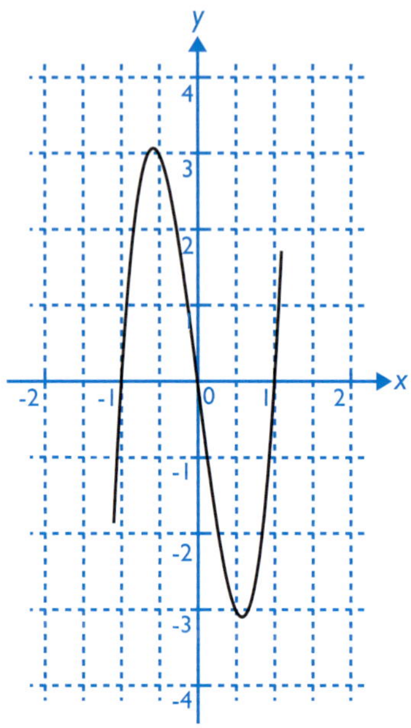

Figure 7.7

b. Use the line you drew for **7a** to approximate $g(0)$, and then draw a second tangent line from the new point to get a second approximation to $g(0.5)$. Use your calculator to find the error in each approximation.

8. **a.** Use Euler's Method with three steps and $\Delta x = 0.5$ to approximate the solution to the differential equation $\frac{dy}{dx} = 1 - 0.2x^2 - y^2$ with starting point (2, 0).

b. Find the equations of the three tangent lines you used for **8a**.

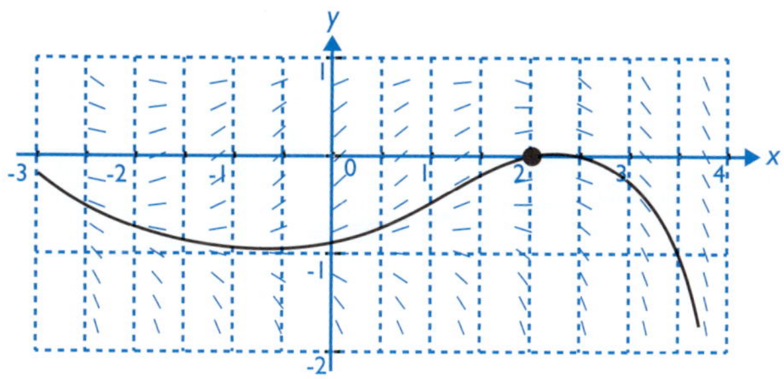

Figure 7.8

c. The solution through (2, 0) is shown on the slope field in Figure 7.8. Draw the three tangent lines you found for **8b** and mark the intermediate points you found in **8a**.

9. **A Surprising Extension** Something interesting happens if we write down an Euler approximation table using formulas for each term instead of values. To be specific, suppose that our starting point is $(a, f(a))$ and that the function we are seeking is $y = f(x)$ with derivative $f'(x)$.

The first two lines of the Euler table are shown in Table 7.2. The first line gives the coordinates of the starting point and the second line gives the tangent line $y = f(a) + f'(a)\Delta x$. The values in the first column have been filled in for you.

a. Complete the rest of the table.

x	y	$\Delta y = f'(x)\Delta x$
a	$f(a)$	$f'(a)\Delta x$
$a + \Delta x$	$f(a) + f'(a)\Delta x$	$f'(a + \Delta x)\Delta x$
$a + 2\Delta x$		
$a + 3\Delta x$		
$a + 4\Delta x$		

Table 7.2

b. Write a formula for the y-value that would be entered on row n.

c. The sum you found in **9b** should look familiar. What is the name for this formula? (If the y-value formulas do not look familiar, then perhaps a slightly different notation will help. Try renaming $a = x_0$ and call the successive x-values $x_1, x_2, x_3, \ldots$. Rewrite your formula in this notation in row x_n.)

d. Show that, as the number of steps n approaches infinity, this sum approaches a definite integral. What theorem does this illustrate?

CHAPTER 8

Review Problems and Self-Assessment

Review Problems

1. On a copy of Figure 8.1, sketch a slope field for $\frac{dy}{dx} = x - y$. Draw tangents at each point with integer coordinates.

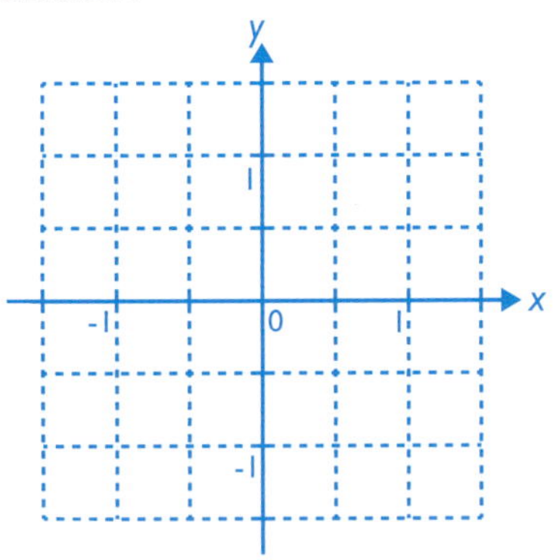

Figure 8.1

2. On a copy of Figure 8.1, sketch a slope field for $\frac{dy}{dx} = \frac{1}{x - y}$. Draw tangents at each point with integer coordinates.

3. Figure 8.2 shows a slope field for some differential equation. Sketch the solution passing through (0, 0).

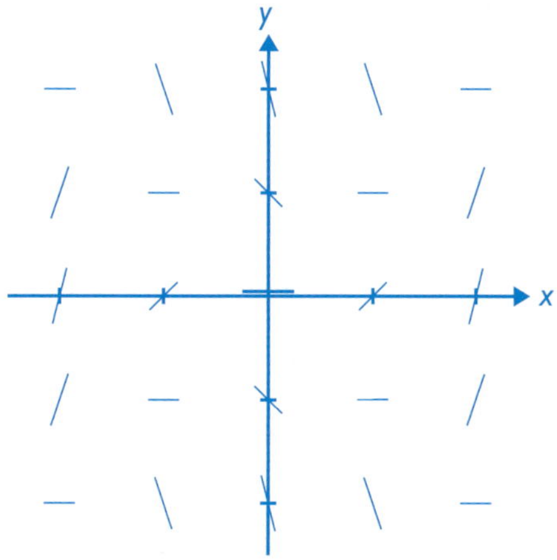

Figure 8.2

4. Figure 8.3 shows a slope field for some differential equation. Sketch the solution passing through (1, −1).

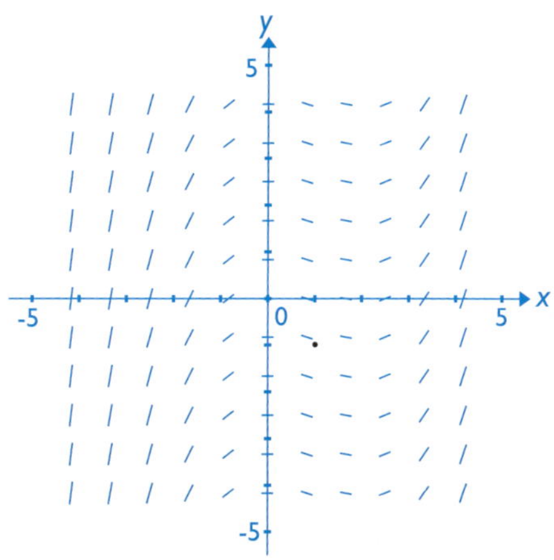

Figure 8.3

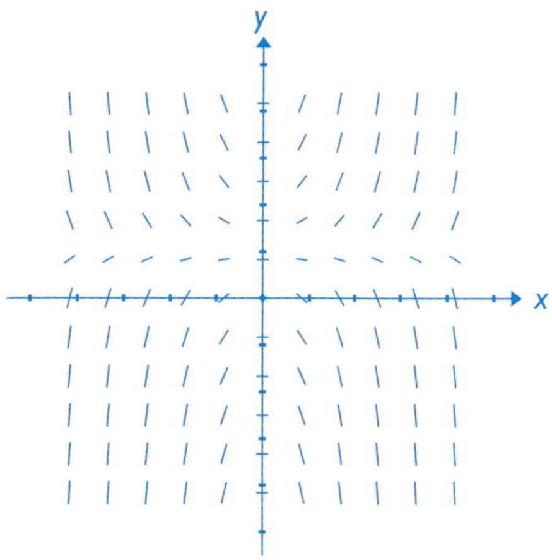

Figure 8.4

5. Why can Figure 8.4 not be a slope field for $\frac{dy}{dx} = xy - 2x$?

6. Figures 8.5a–f show six slope fields. Match them with these six differential equations.

i) $y' = x$ ii) $y' = x - 2$ iii) $y' = x + 1$
iv) $y' = y + 1$ v) $y' = 2y$ vi) $y' = y$

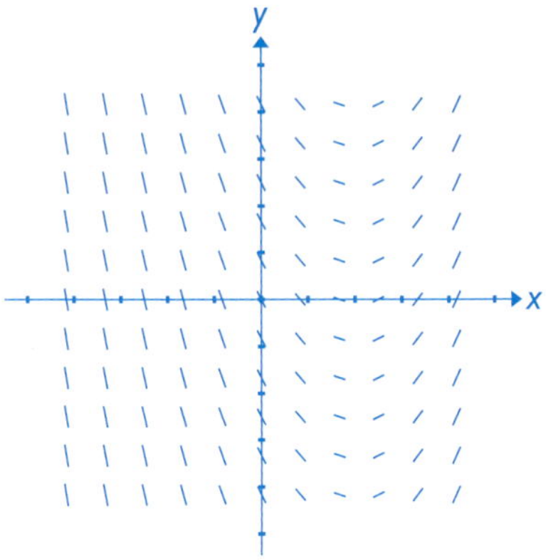

Figure 8.5a

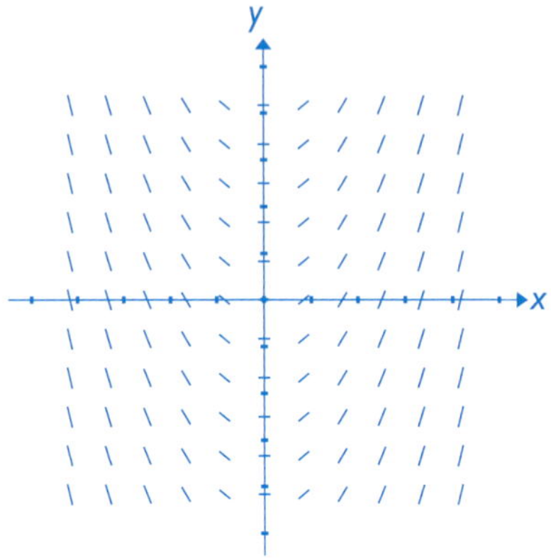

Figure 8.5b

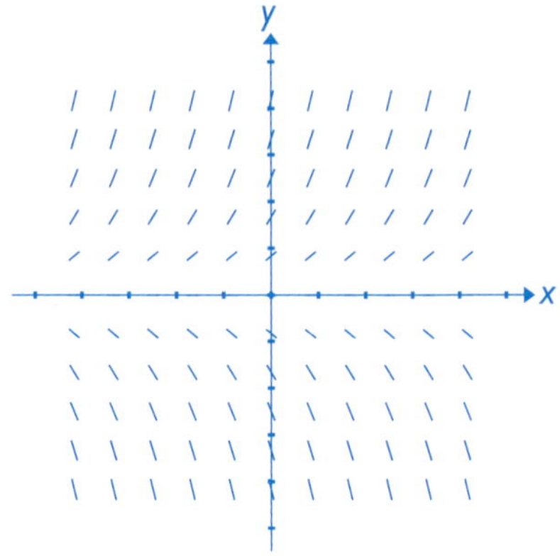

Figure 8.5c

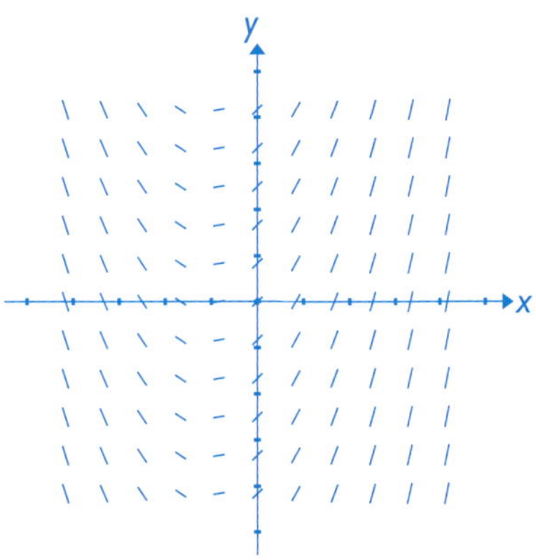

Figure 8.5d

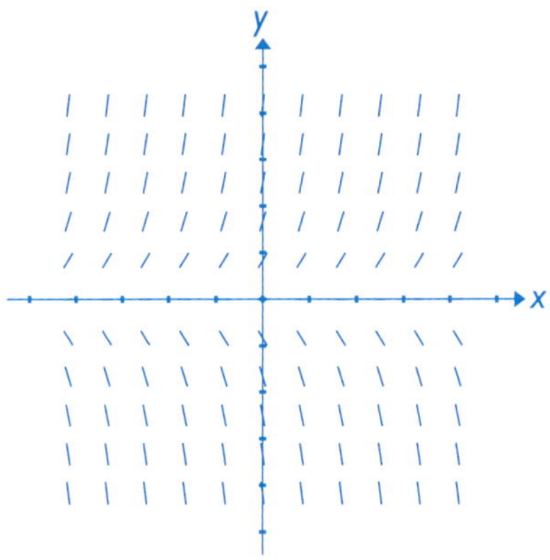

Figure 8.5e

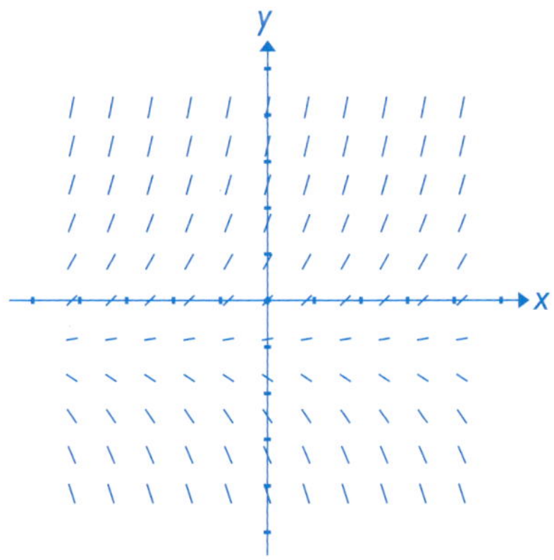

Figure 8.5f

7. Figures 8.6a–f show six slope fields. Match them with these six differential equations.

i) $y' = xe^{-x}$	**ii)** $y' = xe^{-y}$	**iii)** $y' = ye^{-x}$
iv) $y' = ye^{x}$	**v)** $y' = ye^{-y}$	**vi)** $y' = ye^{y}$

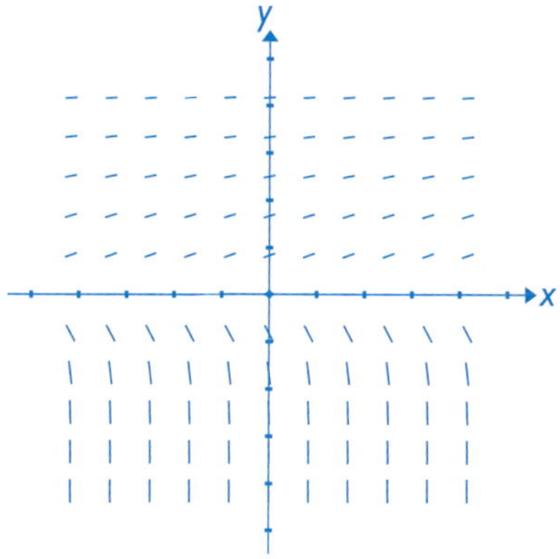

Figure 8.6a

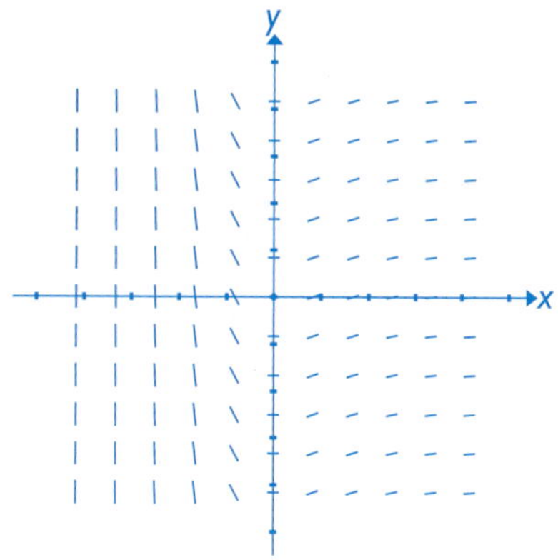

Figure 8.6b

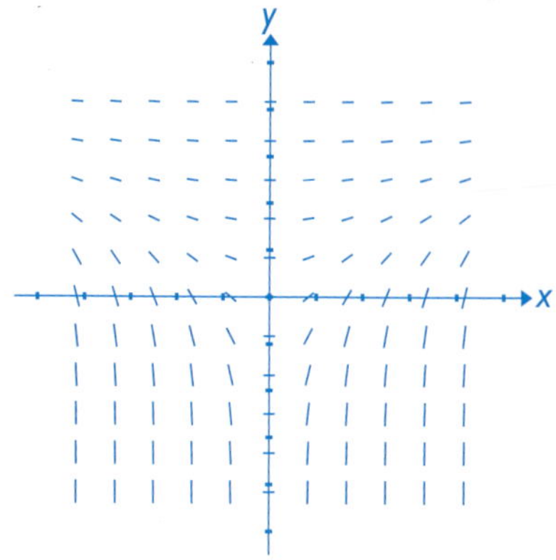

Figure 8.6c

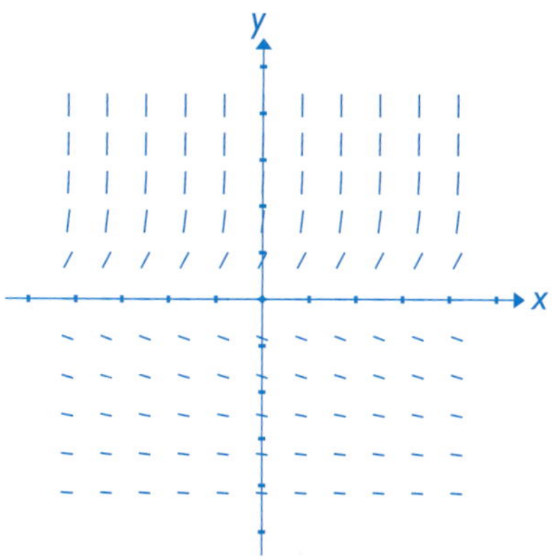

Figure 8.6d

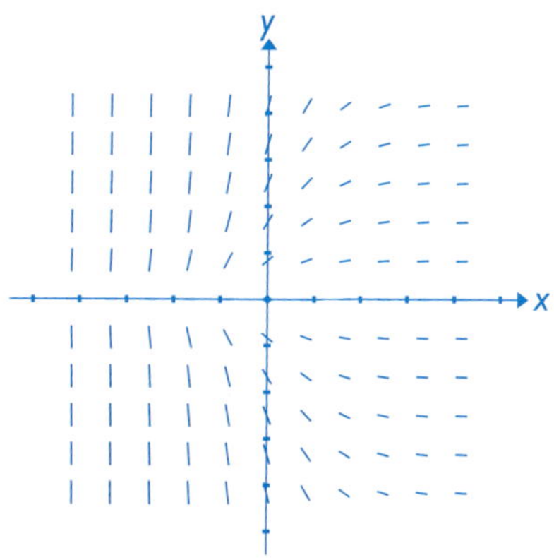

Figure 8.6e

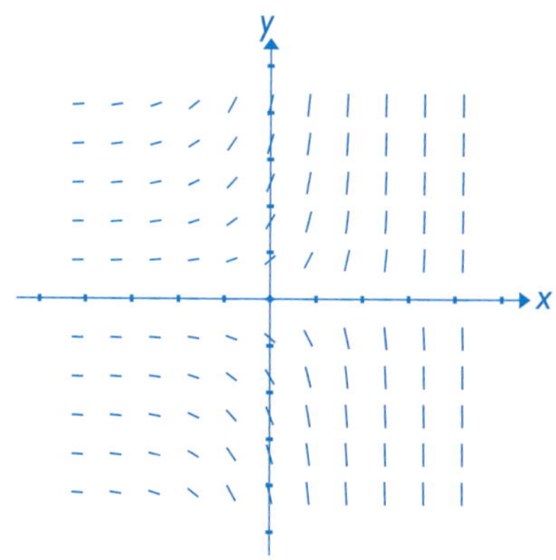

Figure 8.6f

8. What can you say about the symmetry of a slope field for $\frac{dy}{dx} = f(y^2)$?

9. The tangent lines for slope fields for functions of x have parallel tangents along every vertical line. What can you say about the parallel tangent lines of $\frac{dy}{dx} = g(x + y)$?

10. How many variables are involved in the form of the derivative whose slope field is shown in Figure 8.7?

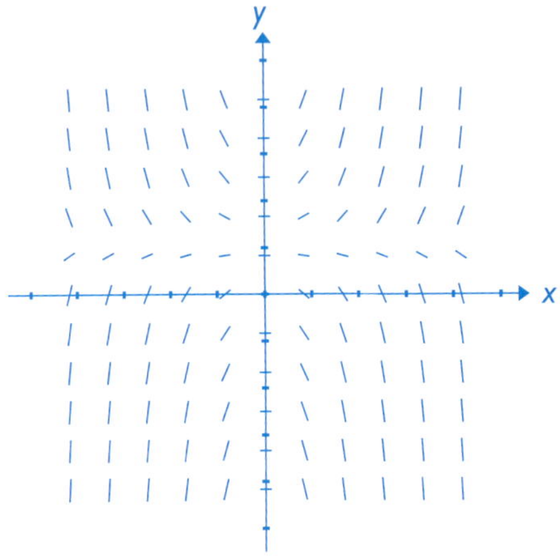

Figure 8.7

11. Sketch the particular solution passing through (1, 2) for the differential equation whose slope field is shown in Figure 8.8.

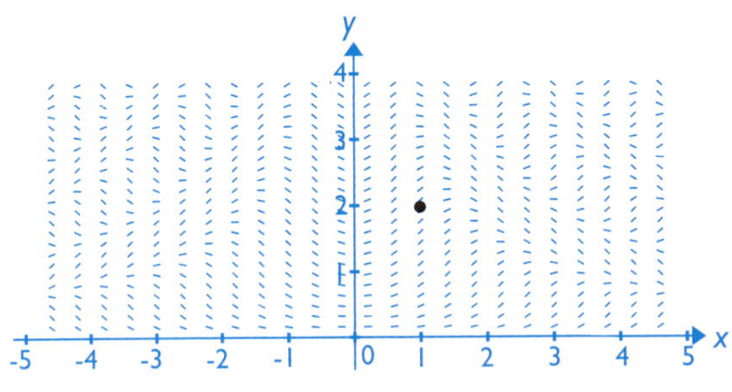

Figure 8.8

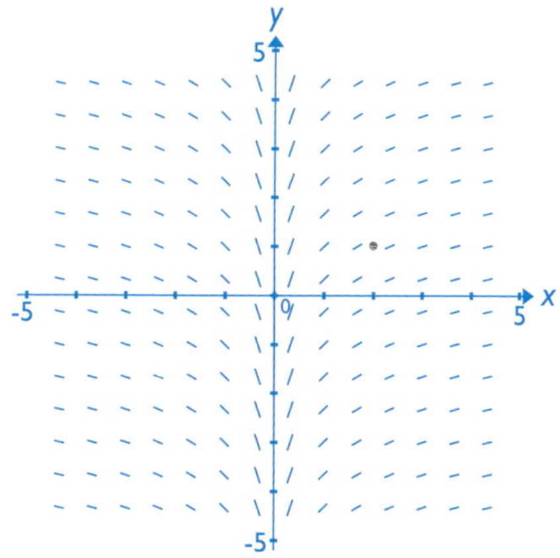

Figure 8.9

12. Sketch the particular solution passing through (2, 1) for the differential equation whose slope field is shown in Figure 8.9.

13. Sketch the particular solution passing through (2, 2) for the differential equation whose slope field is shown in Figure 8.10.

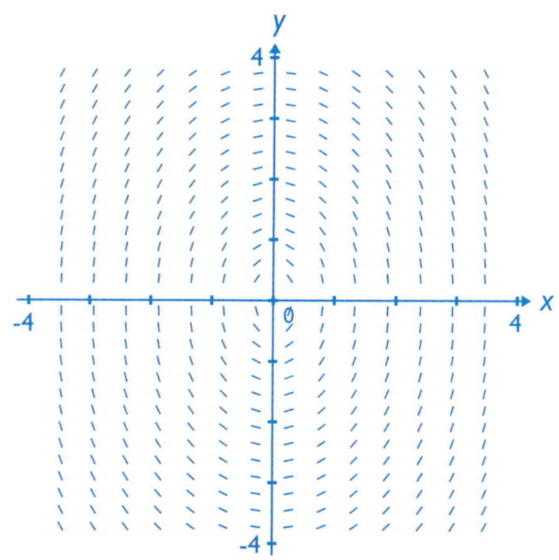

Figure 8.10

14. Find the stable equilibrium solution(s) to $\frac{dy}{dx} = 3y^2 - 7y + 4$.

15. Find the equilibrium solutions to $\frac{dy}{dx} = e^{-2y} - 3e^{-y} - 10$ and decide if each is stable or unstable. If stability depends upon the domain, give the domain.

16. Find the equilibrium solutions to $\frac{dy}{dx} = -0.5(y-2)(y+3)$ and decide if each is stable or unstable. If stability depends upon the domain, give the domain.

17. Show that there is only one linear solution to $\frac{dy}{dx} = (x+y)(y+14)$.

18. Analyze the types of solutions possible for the differential equation whose graph against y is shown in Figure 8.11.

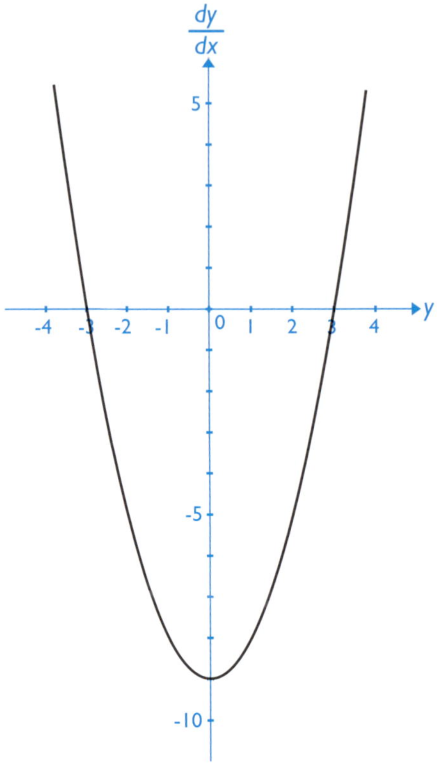

Figure 8.11

19. Use Euler's Method with 5 steps to approximate $y(3.5)$ if $\frac{dy}{dx} = y - 2x$ and the solution passes through $(3, -1)$.

20. Use Euler's Method with 5 steps to approximate $y(4)$ if $\frac{dy}{dx} = \frac{2}{\sqrt{y}}$ and the solution passes through $(2, 4)$.

Self-Assessment

1. On a copy of Figure 8.12 draw a slope field for $\frac{dy}{dx} = x + y$ at points with integer coordinates.

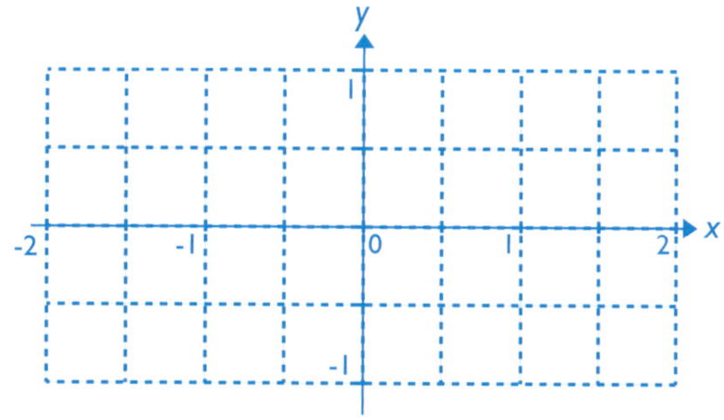

Figure 8.12

2. Match the differential equation to its slope field.

i) $y' = 2x$ **ii)** $y' = \frac{x}{2}$ **iii)** $y' = x^2$ **iv)** $y' = x^{\frac{1}{3}}$

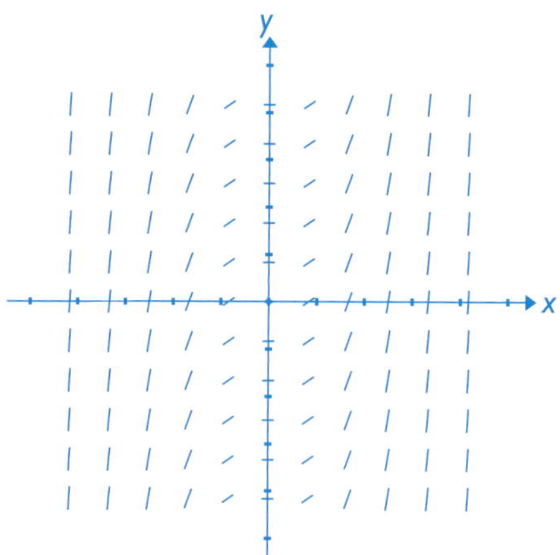

Figure 8.13a

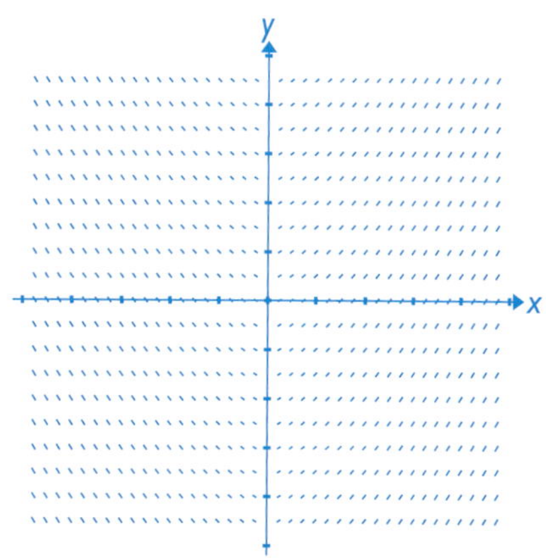

Figure 8.13b

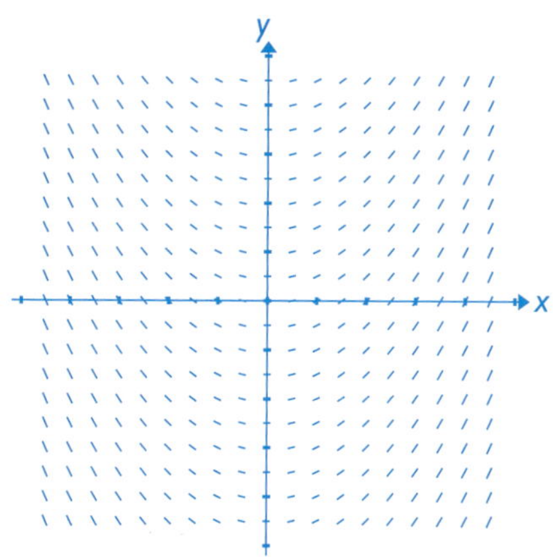

Figure 8.13c

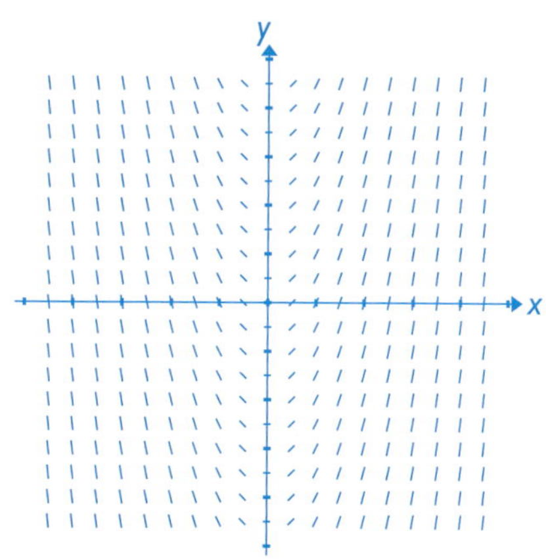

Figure 8.13d

3. Match the differential equation to its slope field.

i) $y' = 1 - xy$ ii) $y' = 1 + xy$ iii) $y' = xy - 1$ iv) $y' = -xy - y$

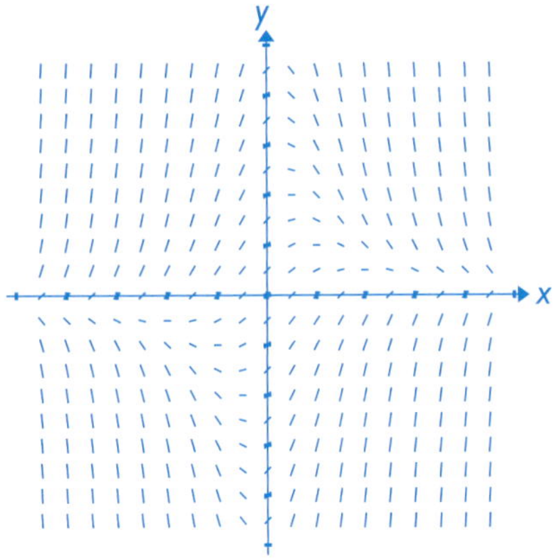

Figure 8.14a

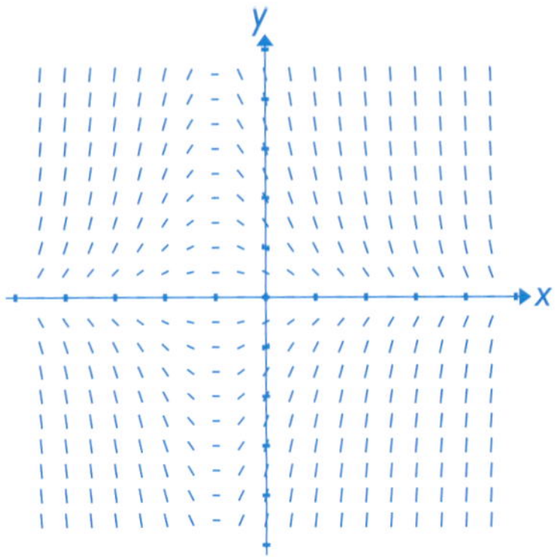

Figure 8.14b

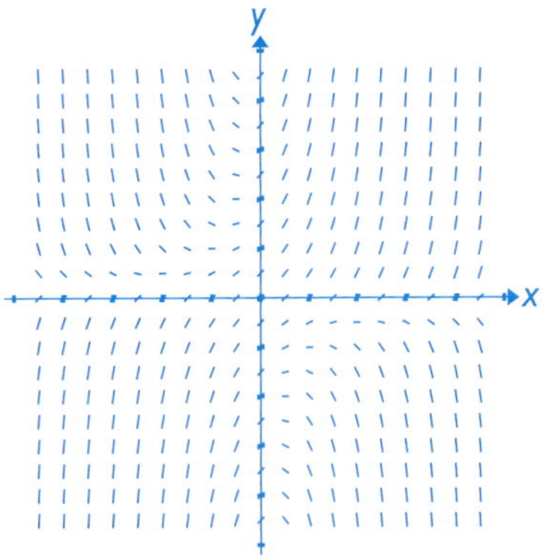

Figure 8.14c

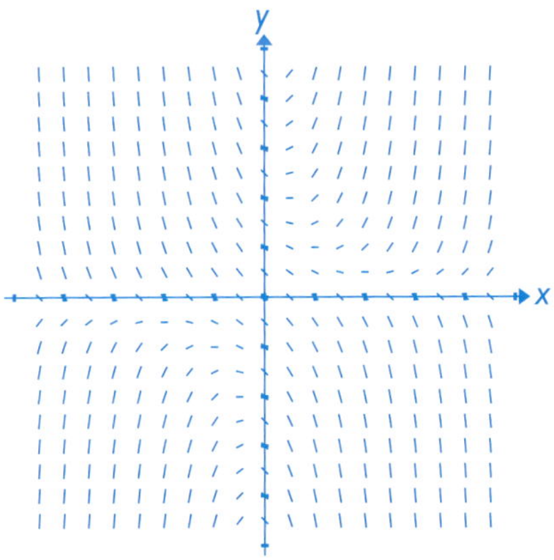

Figure 8.14d

4. Match the differential equation to its slope field.

i) $y' = -xy$ **ii)** $y' = \frac{x}{y}$ **iii)** $y' = 2^{-xy}$ **iv)** $y' = 2^{xy}$

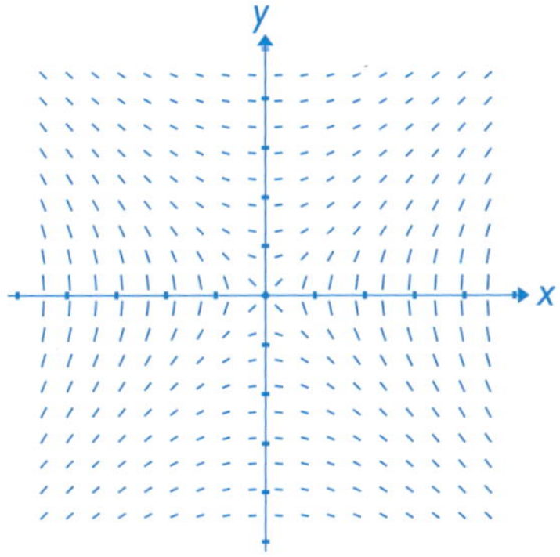

Figure 8.15a

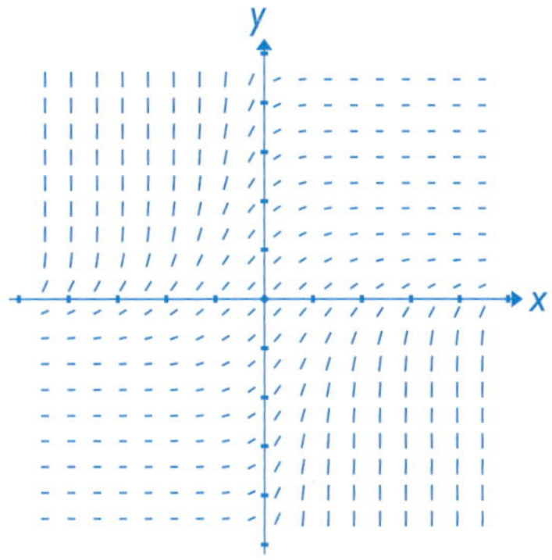

Figure 8.15b

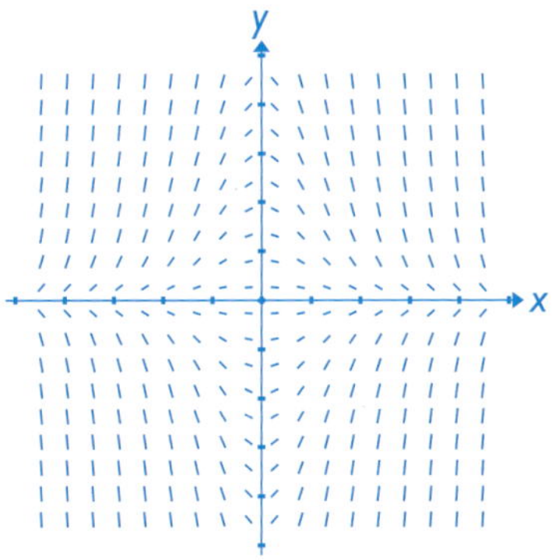

Figure 8.15c

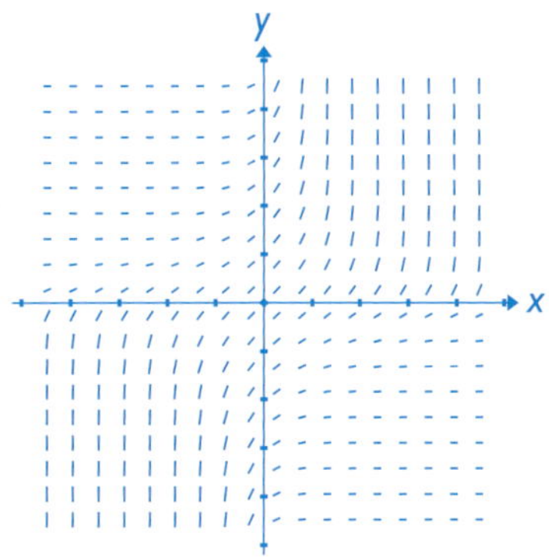

Figure 8.15d

5a. Is the differential equation of the slope field in Figure 8.16 a function of x only, y only, or an implicit function of both x and y?

b. On a copy of the figure sketch the solution passing through (0, 0).

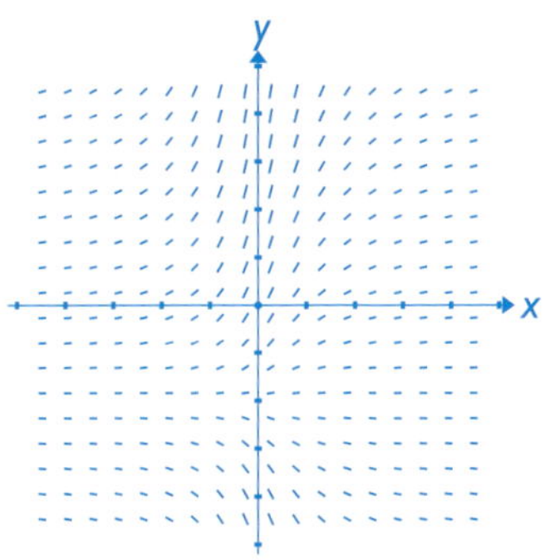

Figure 8.16

6a. Is the differential equation of the slope field in Figure 8.17 a function of x only, y only, or an implicit function of both x and y?

b. On a copy of the slope field in Figure 8.17 sketch the particular solution through (2, 0) and the solution through (2, −4).

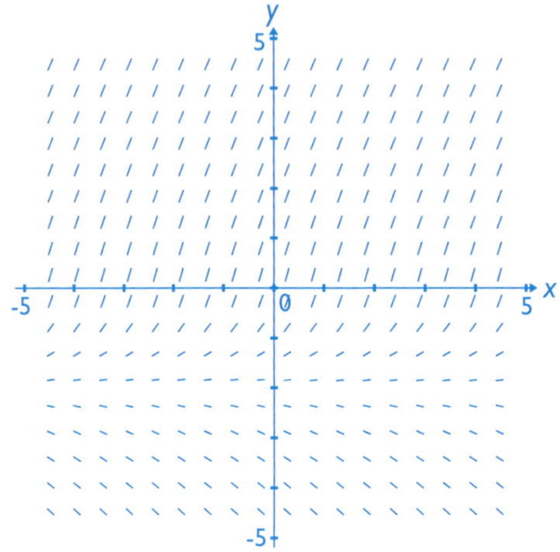

Figure 8.17

7. Find the equilibrium solutions to $\frac{dy}{dx} = y^2 - 4y - 21$ and decide whether each is stable or unstable.

8. Find all linear solutions, if any, to the differential equation $\frac{dy}{dx} = y - 2x + 1$.

9. Figure 8.18 shows a slope field for the implicit differential equation $\frac{dy}{dx} = \frac{x^2 - 2y}{y^2 + 2x + 1}$.

a. On a copy of Figure 8.18 sketch the solutions passing through (0, 9), (−12, −12) and (8, 0).

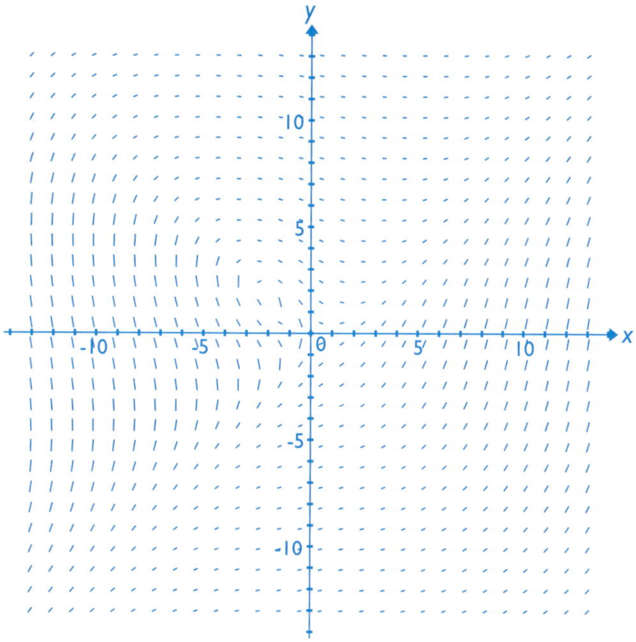

Figure 8.18

b. The slope field appears to have a linear boundary line. Approximate its equation.

c. Find a general formula for the implicit function of which this is the derivative by rewriting the differential equation as $y^2\frac{dy}{dx} + 2x\frac{dy}{dx} + \frac{dy}{dx} - x^2 + 2y = 0$ and then antidifferentiating.

10. Approximate $y(2)$ using Euler's Method with $\Delta x = 0.1$ if $\frac{dy}{dx} = y - 2x$ and the solution passes through (1.6, 2).

APPENDIX A

Slope Fields with the TI-83 and TI-84 Calculators

This is a program for the TI-83 and TI-84 series of graphing calculators. It creates a slope field. Before running the program enter the derivative on the "$y=$" screen. Specific instructions follow the program.

The calculator instructions appear on the left. The comments on the right should help you find the correct keys to press on the calculator keyboard.

Program: Slopefield

:ClrDraw	Clears the graphics screen. 2nd-Draw>1
:FnOff	Turns off any graphs that may be entered. Vars>y-Vars>4
:Input "Nx:",S	Asks for the number of tangent lines to plot across the screen; *"input" is found under Prgm>I/O>1.*
:(Xmax-Xmin)/S→ A	Computes distance between computed points. Use STO for →.
:A/4→ U	
:Input "NY:",T	Asks for the number of tangent lines to plot up & down.
:(Ymax-Ymin)/T→ B	Computes y-distance between computed points.
:B/4→ V	
:For (X, Xmin, Xmax, A)	Computes X values from Xmin to Xmax in steps A. Pgm>4.
:For (Y,Ymin,Ymax, B)	Computes Y values from Ymin to Ymax in steps B.
:If Y_2=0	Test if the derivative fails to exist. Use vars>y-vars>4.
:Then	"Then" is at Prgm>2.
:Line(X,Y-V,X,Y+V)	If so, draw a vertical tangent line. "Line" is at 2nd-Draw > 2

Program	Explanation
:Y1/Y2→ M	Compute slope of tangent.
:If abs(MU)<V	Test steepness of tangent to keep line segment small.
:Then	Find "abs(" on Math>Num>1. Find "<" on 2nd-Test>6
:Line(X-U,Y-MU, X+U,Y+MU)	
:Else	Prg>3
:Line(X-V/M,Y-V, X+V/M,Y+V)	
:End	Ends the "if-then-else" test. End is at Prg>7
:End	Ends the For (Y process).
:End	Ends the For (X process).
:End	Ends the program.

Explanation of the Program Logic

The program counts from the left-edge of the screen (*X*min) to the right edge (*X*max) and from the bottom of the screen (*Y*min) to the top (*Y*max). The number of horizontal and vertical lines are input and the distance between each point is computed. These two loops determine the points (X, Y) at which tangent lines are drawn.

To keep the tangent lines from overlapping, they are drawn to fit a box $A/4$ units across and $B/4$ units up and down, centered on (X, Y).

At each point a tangent line is drawn. First Y_2 is tested. If it is zero, a vertical tangent line is drawn. If it is not zero, then the tangent has finite slope $M = \frac{Y_1}{Y_2}$. If the slope is not too steep, the line is drawn from the left edge of the box to the right edge. If the line is too steep, then it is drawn from the bottom of the box to the top.

Using the Program

The derivative must be expressed explicitly as a function of X and Y. Enter the numerator of the derivative in Y_1 and the denominator in Y_2. If the derivative is not written as a fraction then enter $Y_2 = 1$.

Select an appropriate drawing window and run the program. When prompted, enter the number of lines across the screen (NX) and the number of lines up and down the screen (NY).

Test the program with $NX = 9$, $NY = 9$, $y' = X$ in a window $[-4.5, 4.5]$ H $[-4.5, 4.5]$. The slope field should look like the given figure.

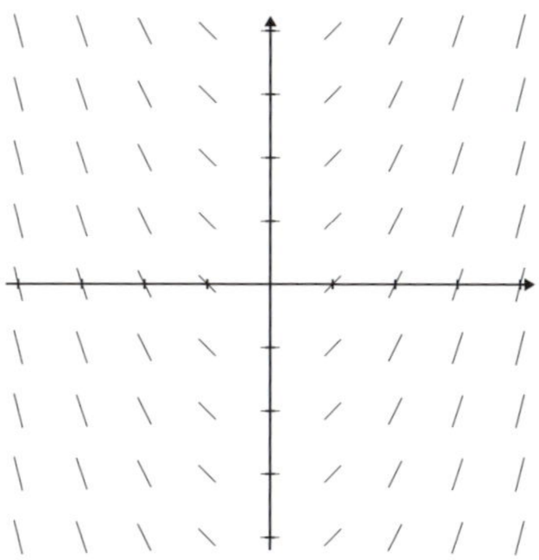

APPENDIX B

Euler's Method with the TI-83 and TI-84 Calculators

This is a program for the TI-83 and TI-84 series of graphing calculators. It uses the calculator's list features to generate approximate solutions to differential equations using Euler's Method (as presented in Chapter 7). Before running the program enter the derivative on the "$y=$" screen. Specific instructions follow the program.

The calculator instructions appear on the left. The comments on the right should help you find the correct keys to press on the calculator keyboard.

Program: Euler

:ClrList L_1	Stat>4. L_1 holds X-values, L_2 holds Y-values, L_3 holds ΔY values. ClrList empties the list of any old data.
:ClrList L_2	
:ClrList L_3	
:	
:Disp "INITIAL X"	Prgm I/O>3. Creates on screen prompts.
:Input X	Prgm I/O>1. Stores input data in variables X, Y, F, D.
:Disp "INITIAL Y"	
:Input Y	
:Disp "FINAL X"	
:Input F	
:Disp "FINAL Y"	
:Input D	
:	
:1→M	M counts through the iterations. Initially $M = 1$. Use "STO" for →.
:X→L_1(M)	Stores initial values in L_1 and L_2
:Y→L_2(M)	

```
:If y2 = 0
:Then
:Goto 0
:Else
:y1D/y2→ L3(M)
:END
```

If, Then, Else are on Prgm> Ctl menu. This tests if the denominator of dy/dx is 0. If it is the program stops. Prgm>Ctl>0

Stores ΔY in list 3.

```
:
:While X<F
:L1(M)+D→ L1(M+1)
:L2(M)+L3(M)→ L2(M+1)
:M+1→ M
:L1(M)→ X
:L2(M)→ Y
```

While is on Prgm> Ctl>5. A<A is on Math>Test. This repeats the commands that follow so long as $X<F$. As soon as X reaches or exceeds F this loop Ends.

The number in L_1 has ΔX added to it and L_2 has ΔY added. Then these numbers are stored back in X and Y so that y_1 and y_2 can compute the next value of $\frac{dy}{dx}$.

```
:
:If y2 = 0
:Then
:Goto 0
:Else
:y1D/y2→ L3(M)
:END
:
Lbl0
:END
```

Lbl is on Prgm > Ctl > 9.

Instructions: Enter the numerator of $\frac{dy}{dx}$ in y_1 and the denominator of $\frac{dy}{dx}$ in y_2. If necessary make $y_2 = 1$, but do not leave this blank. When prompted, give the initial point (X, Y), the change in x, ΔX, and the final x-value.

When the program finishes you will find the solution in the first three columns of the list editor. Press Stat > 1 to open the editor. L_1 will hold the sequence of x-values, L_2 will hold the sequence of y-values, and L_3 will hold the sequence of Δy-values.

Test the program by entering $y_1 = x$ and $y_2 = 1$. Run the problem and use the inputs

Initial $X = 0$

Initial $Y = 0$

Delta $X = .5$

Final $X = 2$.

The program should produce the lists:

L_1	L_2	L_3
0	0	0
0.5	0	0.25
1	0.25	0.5
1.5	0.75	0.75
2	1.5	1

The solution is that when $x = 2$, $y \approx 1.5$.

APPENDIX C

Euler's Method with Excel

Euler's Method can easily be implemented using a spreadsheet. Set up your spreadsheet with five columns as shown:

	A	B	C	D	E
1	Delta x	x-Values	y-Values	Derivative Values	Delta y
2	0.1	2	-1		
3					

1. Row 1 contains column headings. Column A holds Δx. Columns B and C hold the x- and y-coordinates of each point. B2 and C2 hold the values of the initial point while other entries in the column will be computed using Euler's Method. Column D holds the value of the derivative at each point specified in columns B and C. Finally, Column E holds the value of Δy. This is the value computed using a tangent line approximation.

2. In A2, enter the value of Δx; the value 0.1 is illustrated above.

3. In B2 and C2, enter the initial point (x, y). The initial point $(2, -1)$ is shown above.

4. In D2, enter a formula for the derivative using B2 for x and C2 for y. To enter the formula $\frac{dy}{dx} = 1 - x^2 + 2y$, for example, enter $+1 - B2$^2+2*C2 in position D2.

5. In E2, enter +D2*A2. This is the approximation formula($y = y'(x))x$.

To complete the spreadsheet, the formulas and values must be copied down each column. Here is how:

6. In Row 3, enter the following:
   i) In A3, enter +A2
   ii) In B3, enter +B2+A2. (That is, add x to the current x-value to get the next x-value)
   iii) In C3, enter +C2+E2. (That is, add y to the current y-value to get the next y-value)
   iv) In D3, copy the formula from cell D2.
   v) In E3, copy the formula from cell E3.

7. Now copy row 3 into rows 4, 5, 6, . . . as many times as you wish until you reach the final x-value.

Answers to Odd-Numbered Problems and Self-Assessment

Chapter 1

1. a. One.

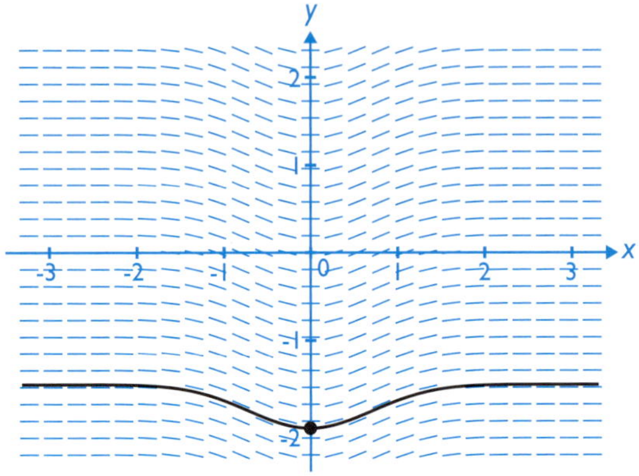

Figure S1.1

b. Figure S1.1 shows the solution through $(0, -2)$. The turning points appear to lie on the y-axis and the solutions to have two points of inflection.

c. The general solution is $y = -\frac{1}{2}e^{-x^2} + C$ and the particular solution through $(0, -2)$ is $y = -\frac{1}{2}e^{-x^2} - \frac{3}{2}$, which has a minimum at $(0, -2)$ and points of inflection at $x = \pm\frac{1}{\sqrt{2}}$.

3. a. One type of solution, all even.

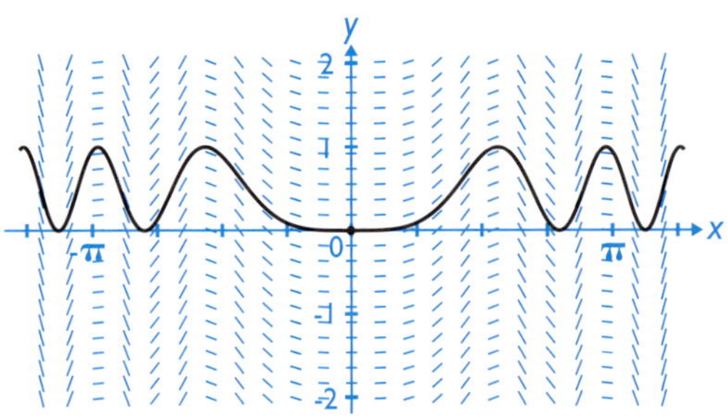

Figure S1.3

b. The general solution is $y = -\frac{1}{2}\cos(x^2) + C$. Figure S1.3 shows the particular solution passing through $(0, 0)$.

5. a. The solution through the origin is odd; all other solutions are neither even nor odd. The solutions appear to have a point of inflection on the y-axis and to increase everywhere.

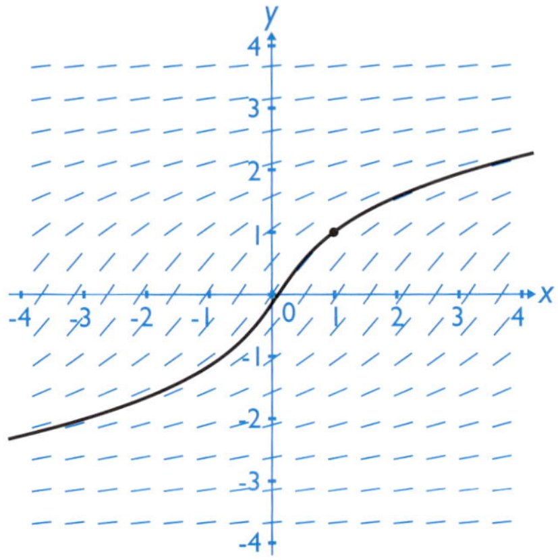

Figure S1.5

b. The general solution is $y = 1.5\tan^{-1}(x) + C$. Figure S1.5 gives the particular solution passing through $(1, 1)$, which does confirm **5a**.

7. a. There appear to be two types of solutions, one on $x < 1$ and one on $x > 1$.

b. There appears to be a vertical asymptote at $x = 1$, which makes sense: $\frac{dy}{dx} = \frac{x^2}{x - x^2}$, so $|y'| \to \pm\infty$ as $x \to 1$.

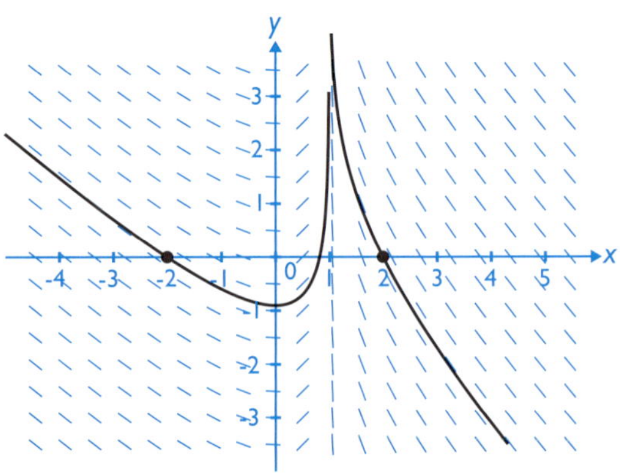

Figure S1.7

c. The general solution is

$$y = -\ln|1 - x| - x + C.$$

The solutions through (2, 0) and (−2, 0) are shown in Figure S1.7 and they do confirm **7a** and **7b**.

9. Figure S1.9 shows three types of solutions. Two are concave down and one is concave up; all have symmetrical finite domains. The solutions shown pass through (0, 4), (0, 2), and (0, −3).

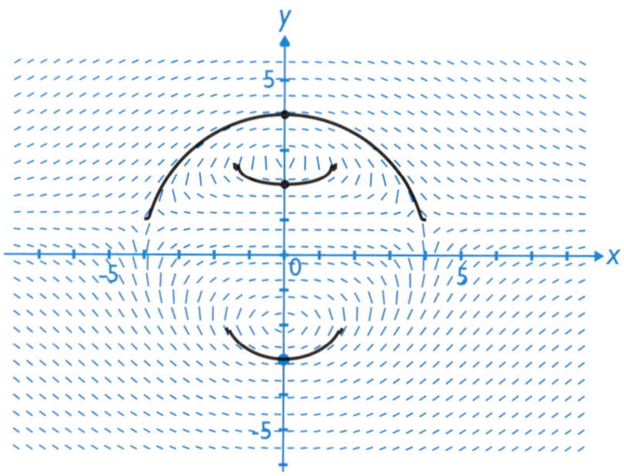

Figure S1.9

11. a. Four types of solutions, one in each quadrant.

b. $(xy)' = C$

Chapter 2

1.

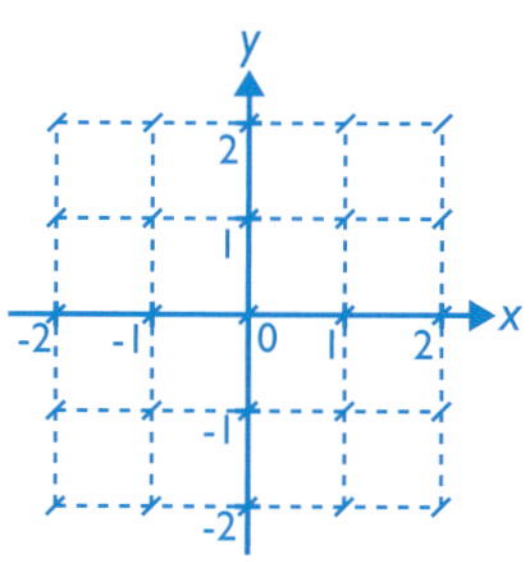

Figure S2.1

3.

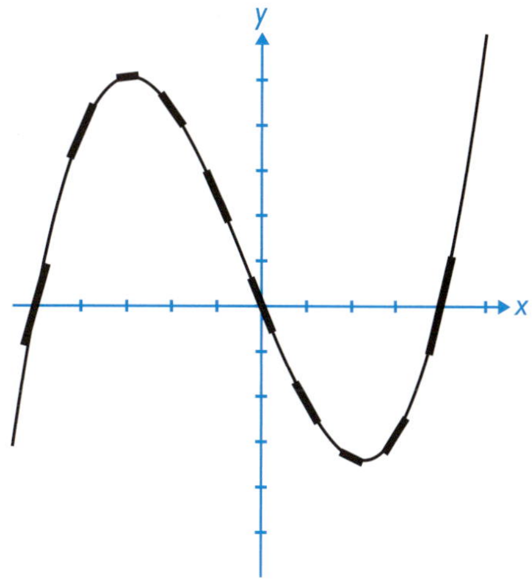

Figure S2.3

5. The slopes should be negative in the second quadrant because $-yx^2 < 0$.

7.

i	ii	iii	iv	v	vi
c	e	a	d	f	b

9.

i	ii	iii	iv	v	vi
a	f	b	e	c	d

11. a. $\frac{1}{2}$

b. Along the diagonal $y = x$ the slopes are undefined, as the tangent lines become steeper and steeper until vertical. Therefore, the derivative does not exist on $y = x$.

13. Because the derivative is a function of y alone, the slopes at each point are independent of x and so the tangent lines are parallel along each horizontal line $y = c$.

15. Because the tangent lines appear to be parallel along each horizontal line $y = c$, the derivative must be independent of x and so is a function of y only.

17. $y' = -1$.

19. The slope fields will be symmetrical through the origin.

Chapter 3

1. a. Figure S3.1 shows a sketch of the solution passing through (0, 2). On that curve, $z(2) \approx 3$.

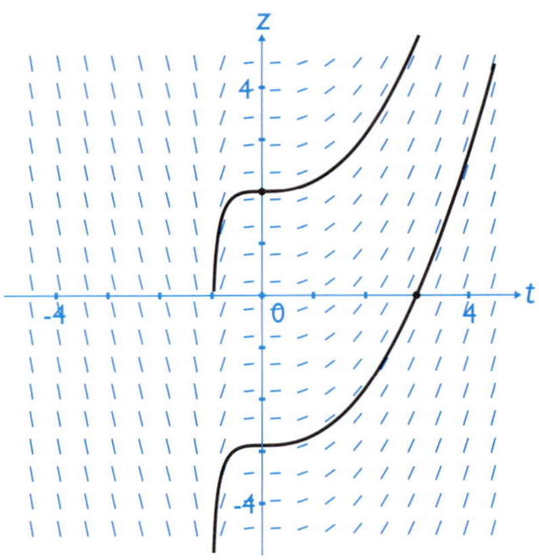

Figure S3.1

b. $z = 0.5t^2 - t + \ln|t + 1| + 2$

and $z(2) \approx 3.099$

3. a. Figure S3.3 shows the solution through $(\pi, 1)$ where $y(\frac{\pi}{2}) \approx 1.25$.

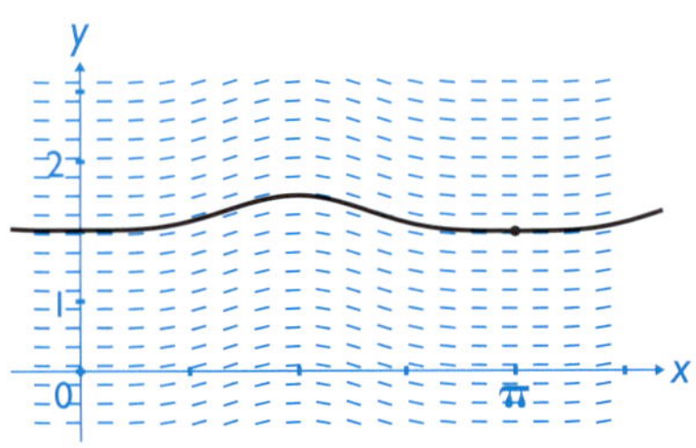

Figure S3.3

b. $y = \frac{1}{4}\sin^4(x) + 1$ and so $y(\frac{\pi}{2}) = \frac{5}{4}$.

5. Figure S3.5 shows the solutions through $(1, -2)$ and through $(1, 3)$.

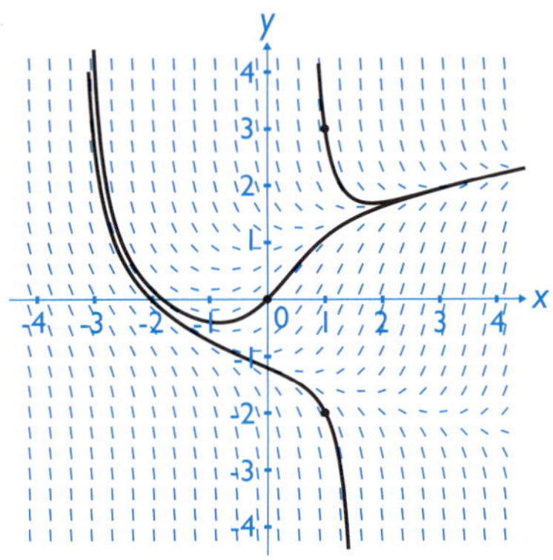

Figure S3.5

a. $y(-2) \approx 0$.

b. $y(3) \approx 2$.

7. a. Substitute $y = -1 - x$ into the differential equation to get $-1 = x + (-1 - x) = -1$.

b. Substitute $y = mx + b$. This gives

$$m = x + (mx + b) = (m + 1)x + b.$$

This equation should hold for all values of x, in particular for $x = 1$. Therefore, $m = m + 1 + b$ so that $b = -1$. The equation should hold for $x = 0$ as well and so $m = b = -1$.

c. If (a, b) lies below $y = -1 - x$ then $b < -1 - a$, so $y' = a + b$ is negative and the solution is decreasing. Therefore, any point on the solution with $x < a$ must have $y > b$ and yet still lie below $y = -1 - x$. Such a point must be closer to the boundary than (a, b).

d. Because solutions that lie below the boundary are always decreasing as x increases, they must increase as x decreases. However, these solutions are bounded above by the boundary and so cannot increase beyond that boundary. They might either level off to a line parallel to the boundary or level off at the boundary. Because the boundary line is the only linear solution, only the second case can occur. That is, the line $y = -1 - x$ is an asymptote to solutions that lie below it.

9. a. If the boundary line were $x + y = 6$ then $y = 6 - x$ would be a solution. If we substitute into the differential equation, we get $-1 = x + y - 6 = 0$. So this line is not the equation of a solution.

b. Substitute $y = mx + b$ in the equation. This gives $m = x + (mx + b) - 6$ or $m = (m + 1)x + b - m - 6$. Because this must hold for all values of x, it must hold when $x = 1$, so $m = b - 5$ and when $x = 0$ so $m = b - m - 6$. Therefore, $m = -1$ and $b = 4$. The boundary line is $y = 4 - x$.

c. There are two types of solutions because no solution can intersect $y = 4 - x$. Therefore, there are solutions below $y = 4 - x$ and solutions above $y = 4 - x$.

11. Figure S3.11 shows the solution through (2, 0).

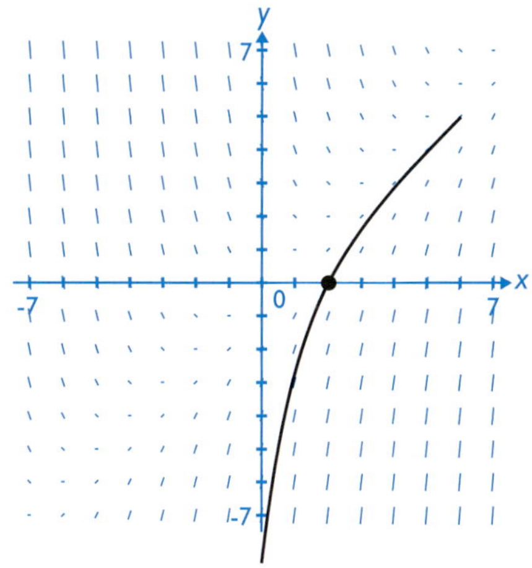

Figure S3.11

a. The solution through (2, 0) increases everywhere, but it is difficult to identify its domain. It appears to have a vertical asymptote. If you see that, then the domain on which it increases is, roughly, $x > 0$.

b. There appears to be a boundary line through (1, 0) and (0, −1), that is, $y = x - 1$. Solutions below this boundary are concave down.

Chapter 4

1.

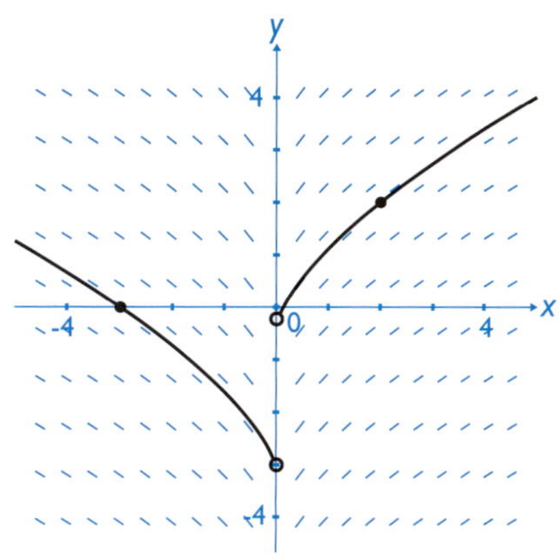

Figure S4.1

3.

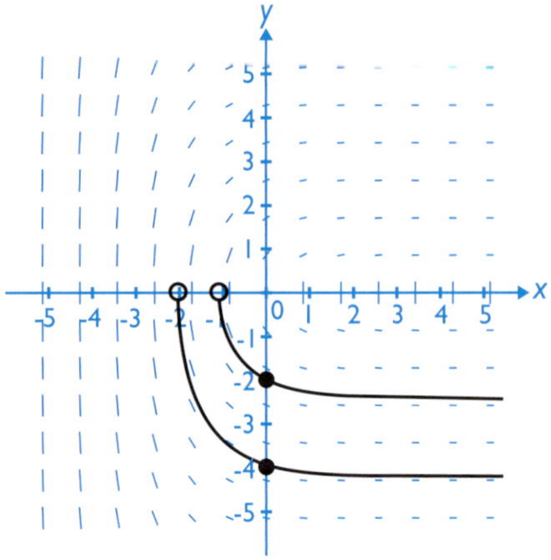

Figure S4.3

5.

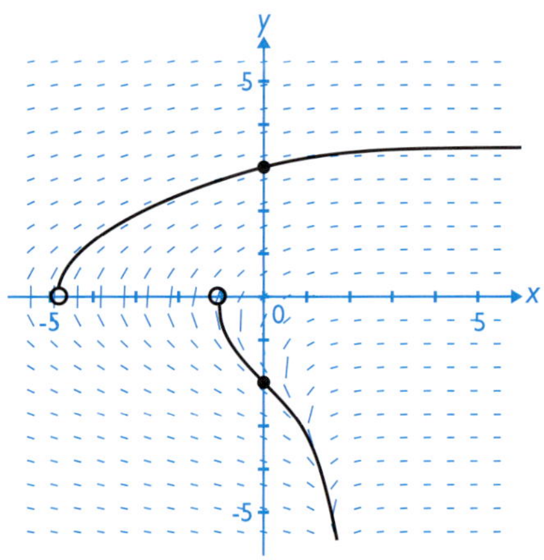

Figure S4.5

7.

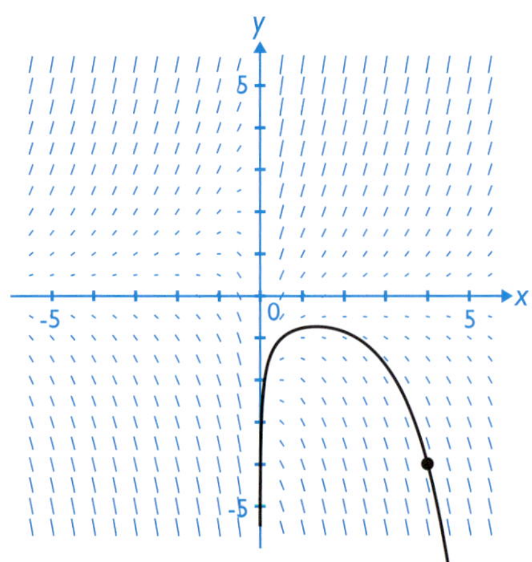

Figure S4.7

9.

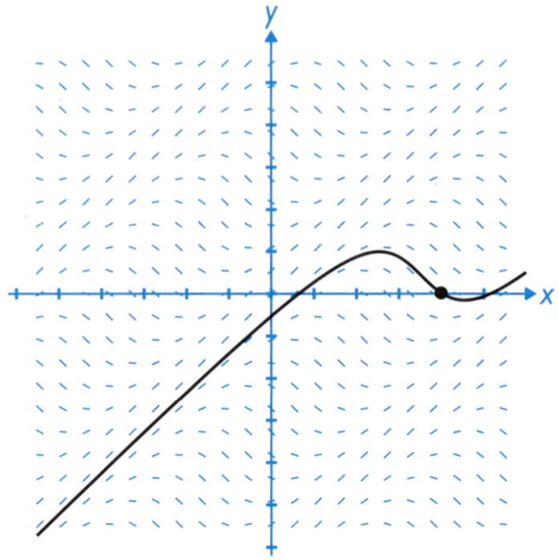

Figure S4.9

11. a.

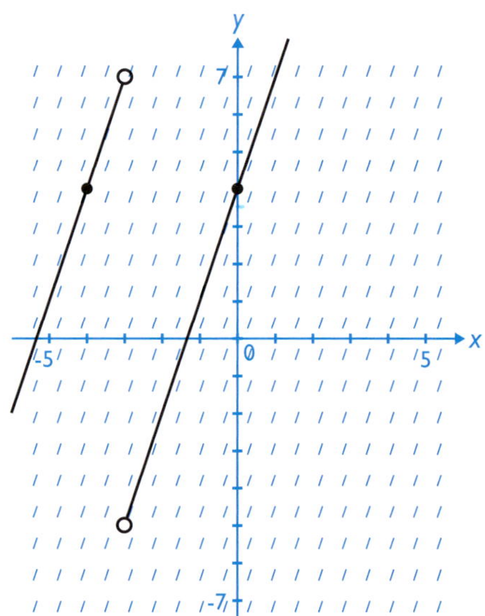

Figure S4.11

b. $\frac{dy}{dx} = \frac{3(x+3)}{x+3} = 3$ when $x \neq -3$, so the general solution is $y = 3x + C$. The particular solutions are **(i)** $y = 3x + 4$ through (0, 4), and **(ii)** $y = 3x + 16$ through $(-4, 4)$. The domain and range of **(i)** are $x > -3$ and $y > -5$. The domain and range of **(ii)** are $x < -3$ and $y < 7$.

Chapter 5

1. There is an unstable equilibrium at $y = \sqrt{8}$ and a stable one at $y = -\sqrt{8}$.

3.

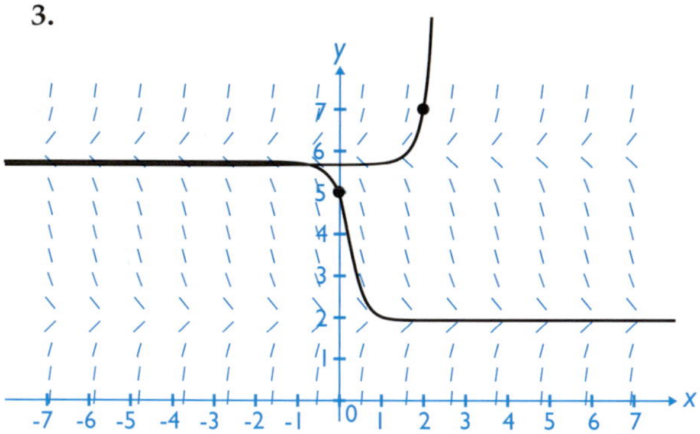

Figure S5.3

5. The differential equation has a stable equilibrium because it approaches $y = 46$ as x approaches infinity.

7. There is no equilibrium solution because f does not have a horizontal asymptote.

9. There are no equilibrium solutions.

11. a. The equilibrium solutions are $y = 2$ and $y = 7$.

b. $\frac{d^2y}{dx^2} = (-2y + 9)\frac{dy}{dx}$. There are points of inflection along $y = 4.5$.

c. Initial values in the range $2 < y < 7$ will have points of inflection along $y = 4.5$.

13. Substitute $y = -x - 3$ into the differential equation to get $\frac{dy}{dx} = -1$, so that $y = -x + C$. If (a, b) is an initial value on $y = -x - 3$ then the particular solution is $b = -a - 3$ and $C = -3$. Therefore, $y = -x - 3$ is a solution.

Chapter 6

1. Figure S6.1 shows that the solutions are always increasing, concave down when $y < 0$ and concave up when $y > 0$, and have a point of inflection on $y = 0$.

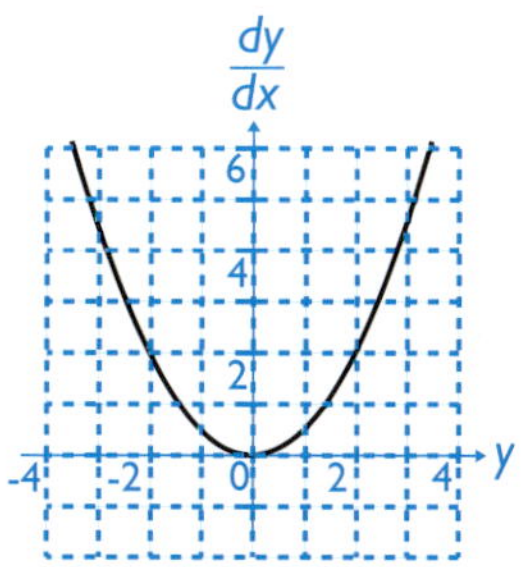

Figure S6.1

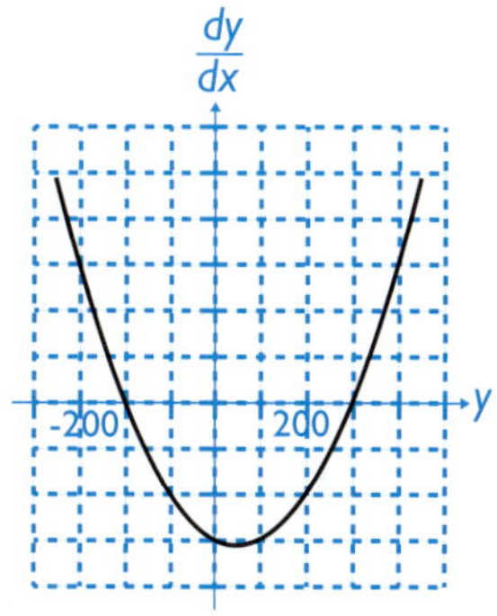

Figure S6.3

3. Figure S6.3 shows that the derivative is positive and decreasing when $y < -200$, so solutions in this range will be increasing and concave down. When $y > 300$, the derivative is positive and increasing, so the solutions in this range are increasing and concave up. Finally, solutions in the range $-200 < y < 300$ are decreasing and change concavity along the horizontal line $y = 50$.

5. Do not be misled by this one. Here the derivative is a function of x, not y, and so the horizontal axis is describing x-values in Figure S6.5. There is only one type of solution curve, a fourth degree polynomial, which has relative minima at $x = -5$ and $x = 3$, where the derivative changes sign from negative to positive, and has a relative maximum at $x = 0$. The solution has points of inflection where $\frac{d^2y}{dx^2} = 3x^2 + 4x - 15$ changes sign, at $x = \frac{5}{3}$ and $x = -3$.

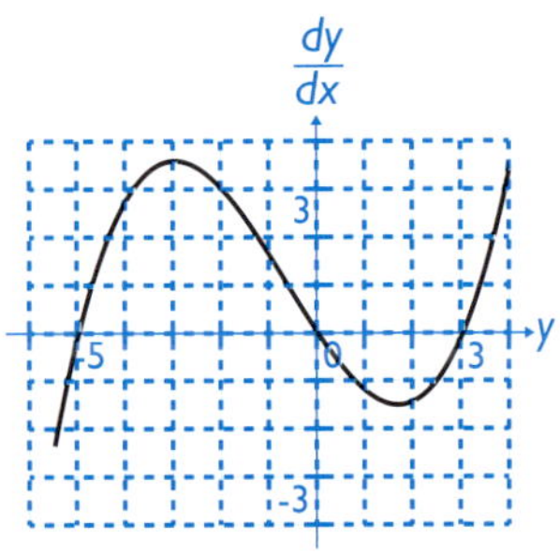

Figure S6.5

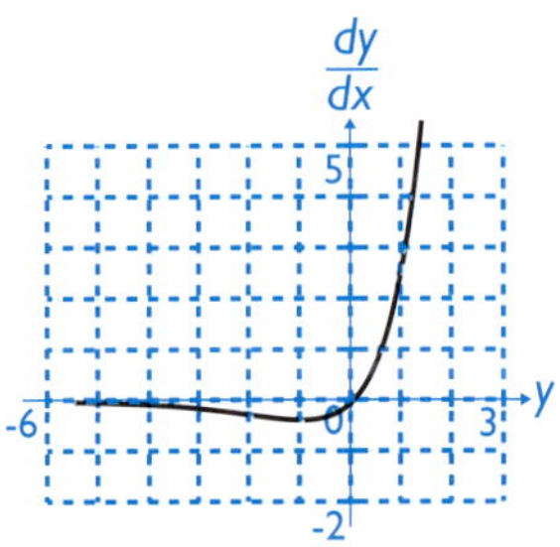

Figure S6.7

7. As seen in Figure S6.7, this derivative is negative for $y < 0$ and positive for $y > 0$. Solutions in the range $y < 0$ will be decreasing functions, while solutions where $y > 0$ are increasing functions.

9. a. The differential equation is

$\frac{dP}{dt} = 0.005P(1200 - P)$. Figure S6.9 shows that solutions with $0 < P < 1200$ are increasing functions that change concavity at $P = 600$. These solutions are concave up until becoming concave down when they hit the line $P = 600$, which marks the time when the deer population is increasing most rapidly.

Solutions with $P < 0$ are concave up and decreasing, but of course the population cannot be negative, so these solutions have no physical interpretation. If $P > 1200$, the solutions are concave up and decreasing.

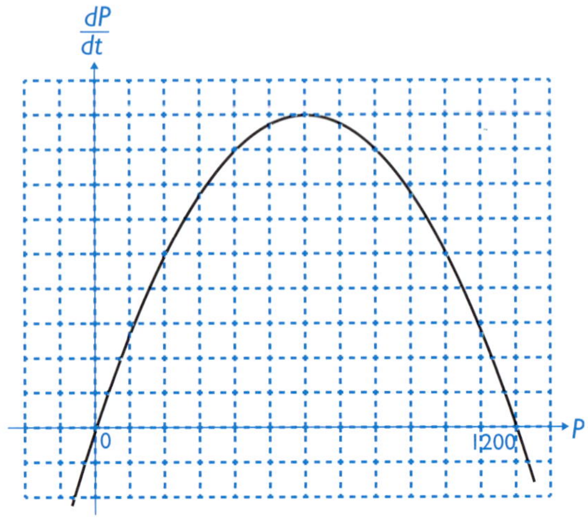

Figure S6.9

b. If the initial value of P is greater than 1200, then the solutions are concave up and decreasing. The meaning is that, if there are more animals than the environment can sustain, the population will die off until it reaches the stable equilibrium of 1200.

11. Differentiate implicitly to get $y'' = 2y\frac{dy}{dx} = 2y^3$. Differentiate again to get $y''' = 6y^4$ and $y^{(iv)} = 24y^5$. It is easy to see the pattern of exponents, but the pattern of coefficients is a little less clear. If you write out the multiplications, without simplifying, you get $y''' = (2)(3)y^4$ and $y^{(iv)} = (2)(3)(4)y^5$. Now it should be clear that $\frac{d^ny}{dx^n} = n!y^{n+1}$ for $n = 1, 2, 3, \ldots$.

13. The equilibrium solutions are $y = -2$ and $y = 4$, so there are three types of solutions.

a. At (0, 0) the slope is -8. The solution through (0, 0) lies between the equilibria, where the derivative is negative. Therefore, it will decrease between the two horizontal asymptotes, reaching its point of inflection at $y = 1$. Figure S6.13 shows a possible sketch.

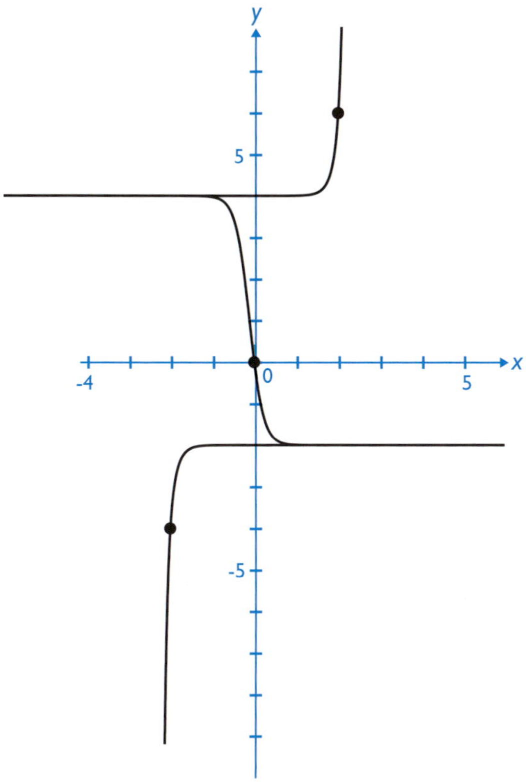

Figure S6.13

b. At (2, 6) the slope is a large positive value. The solution through (2, 6) lies above the equilibrium at $y = 4$. The derivative is positive and increasing, so the solution is concave up and increasing. Figure S6.13 shows a possible solution.

c. At $(-2, -4)$ the derivative is a large positive value. The solution through $(0, -4)$ lies below the equilibrium at $y = -2$. The derivative is positive and decreasing, so the solution is concave down and increasing. Figure S6.13 shows a possible solution.

Chapter 7

1. a.

x	y	$\Delta y \approx y'(x)\Delta x$
2.0	2.0	1
2.5	3	1.25
3.0	4.25	1.5
3.5	5.75	

b. Because $y' = x$, $y = \frac{x^2}{2} + C$, and the solution through (2, 2) is $y = \frac{x^2}{2}$. Therefore, $f(3.5) = 6.125$.

3. a.

x	y	$\Delta y \approx y'(x)\Delta x$
1	1	0.2
1.2	1.2	0.24
1.4	1.44	0.284
1.6	1.724	0.332
1.8	2.056	0.385
2.0	2.441	0.442
2.2	2.883	0.504
2.4	3.387	0.57
2.6	**3.957**	

b. If $\frac{dy}{dx} = \sqrt{xy}$, then $\frac{dy}{\sqrt{y}} = \sqrt{x}\,dx$. Antidifferentiate to get $2\sqrt{y} = \frac{2}{3}x^{\frac{3}{2}} + C$, or $y = (\frac{1}{3}x^{\frac{3}{2}} + C)^2$. Because the solution passes through (1, 1), $C = \frac{2}{3}$. Therefore, $y = (\frac{1}{3}x^{\frac{3}{2}} + \frac{2}{3})^2$ and so $y(2.6) = \mathbf{4.260}$.

5. a. Figure S7.5a shows the solution passing through $(-2, 3)$. Use it to approximate $y(-1) \approx 2.6$.

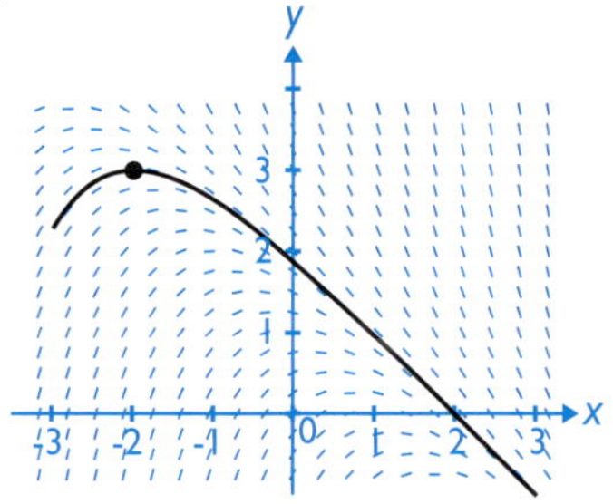

Figure S7.5a

b.

x	y	$\Delta y \approx y'(x)\Delta x$
-2	3	0
-1.5	3	-0.25
-1	2.75	

The approximation is an overestimate because the solution is concave down and so the tangent lines lie above the solution.

c.

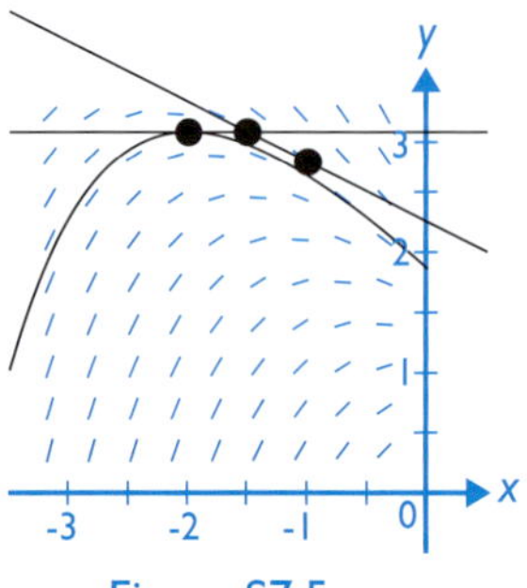

Figure S7.5c

7. a. Figure S7.7 has a tangent line through $(-\frac{1}{2}, 3)$ showing that $g(0.5) \approx 1$.

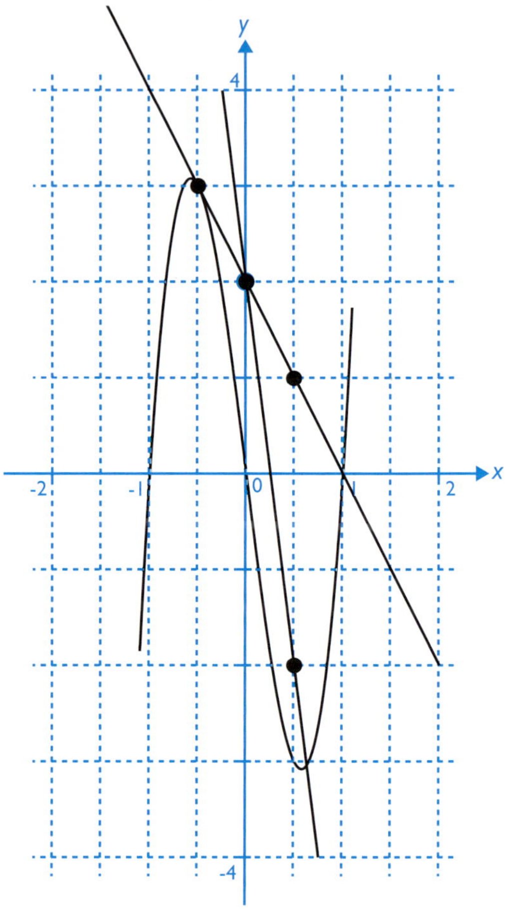

Figure S7.7

b. Figure S7.7 suggests that the y-intercept of the initial tangent line is 2. The figure also shows a second line drawn at that point with slope -8, the slope of $g(x)$ at $x = 0$, and that line suggests that $g(0.5) \approx -2$.

The two-tangent approximation gives a value that is much closer to the correct value. A calculator shows that $g(0.5) = -3$, so the error in a single tangent line approximation of $g(0.5)$ is -4 and the error in a double tangent line approximation is -1.

9. a.

x	y	$\Delta y \approx y'(x)\Delta x$
a	$f(a)$	$f'(a)\Delta x$
$a + \Delta x$	$f(a) + f'(a)\Delta x$	$f'(a+\Delta x)\Delta x$
$a + 2\Delta x$	$f(a) + f'(a)\Delta x + f'(a + \Delta x)\Delta x$	$f'(a+2\Delta x)\Delta x$
$a + 3\Delta x$	$f(a) + f'(a)\Delta x + f'(a + \Delta x)\Delta x + f'(a + 2\Delta x)\Delta x$	$f'(a+3\Delta x)\Delta x$
$a + 4\Delta x$	$f(a) + f'(a)\Delta x + f'(a + \Delta x)\Delta x + f'(a + 2\Delta x)\Delta x + f'(a + 3\Delta x)\Delta x$	$f'(a + 4\Delta x)\Delta x$

b. The approximation for y in row $x = n$ is
$f(a) + f'(a)\Delta x + f'(a + \Delta x)\Delta x +$
$f'(a + 2\Delta x)\Delta x + \ldots + f'(a + (n-1)\Delta x)\Delta x$
or, in summation notation,
$y_n = \sum_{k=0}^{n} f'(a + (k - 1)\Delta x)\Delta x$.

c. If we write $x_n = x_0 + (k - 1)\Delta x$, then the sum is $y_n = \sum_{k=0}^{n} f'(x_{k-1})\Delta x$, a Left Riemann Sum approximation for the function $y = f(x)$ on the interval $[a, b]$.

d. As n approaches infinity, the approximation approaches the definite integral $\int_a^b f(x)dx$.

Chapter 8 Review Problems

1.

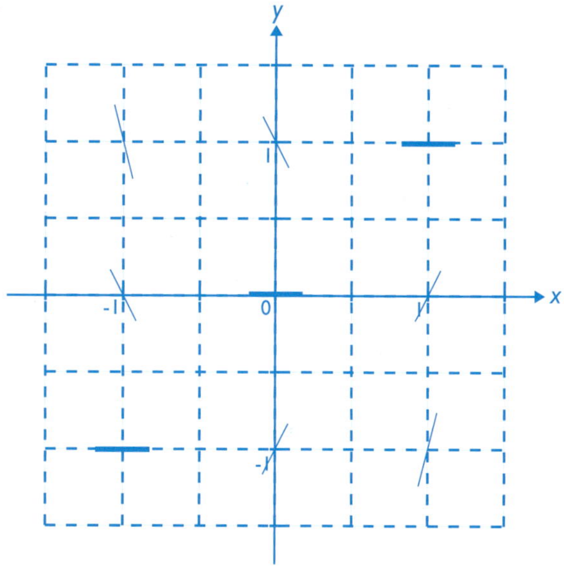

Figure S8.1

3.

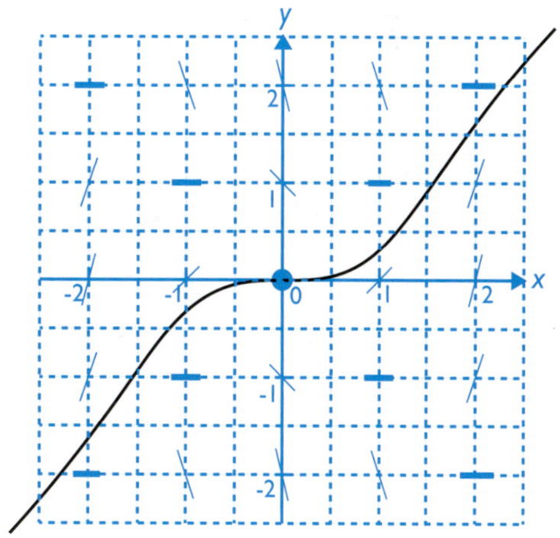

Figure S8.3

5. The slopes should be zero along $y = 2$, not along $y = 1$.

7. **i)** b **ii)** c **iii)** e **iv)** f **v)** a **vi)** d

9. If $x + y = k$, then $\frac{dy}{dx} = g(k)$. That is, tangent lines along any line of the form $x + y = k$ will be parallel.

11.

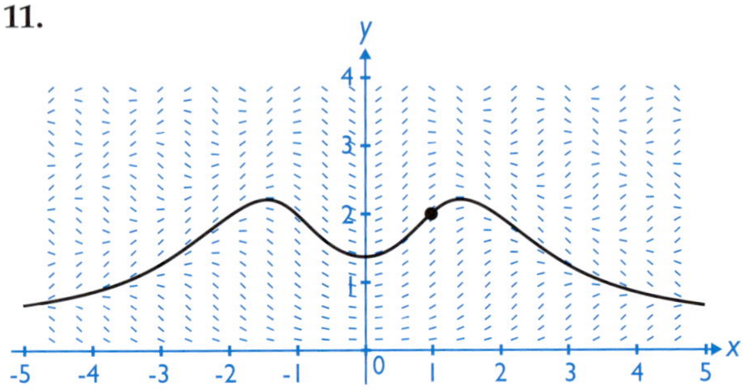

Figure S8.11

13.

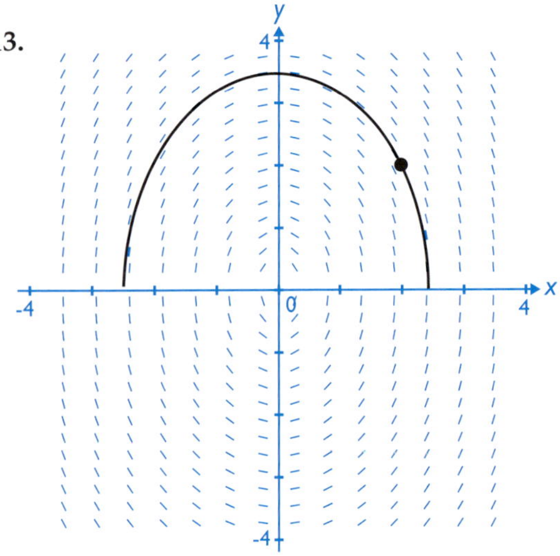

Figure S8.13

15. The derivative is zero when $e^{-y} = 5$, that is, when $y = -\ln(5)$, so this is the only equilibrium solution.

17. To find a linear solution, substitute $y = mx + b$ in the original differential equation. We get $m = (x + mx + b)$ $(mx + b + 14) = ([m + 1]x + b)$ $(mx + [b + 14])$. Multiply out to get $m = m\ [m + 1]\ x^2 + (mb + [m + 1]\ [b + 14])$ $x + b\,[b + 14]$.

The key idea is that this should be an identity, that is, the two sides of the equation should be equal for all values of x. However, the only way the constant on the left can equal the quadratic polynomial on the right is if the coefficients of x^2 and x are zero. That is, $m[m + 1] = 0$ and $(mb + [m + 1]\,[b + 14]\,) = 0$.

There are two possibilities. Either $m = 0$ or $m = -1$. If $m = 0$, then the equation becomes $0 = 0x^2 + (b + 14)$ $x + b(b + 14)$. Because this must be true for all x, the coefficient $b + 14 = 0$, so $b = -14$. That is, $y = -14$. The other option, that $m = -1$, gives $0x^2 - bx + b(b + 14)$. However, this forces $b = 0$, which gives $-1 = 0$, an impossibility. Therefore, the only linear solution is the equilibrium solution $y = -14$.

19.

x	y	$\Delta y \approx y'(x)\Delta x$
3	−1	−0.7
3.1	−1.7	0.48
3.2	−1.22	0.442
3.3	−0.778	0.408
3.4	−0.37	0.377
3.5	0.007	

Complete Answers to Self-Assessment

1.

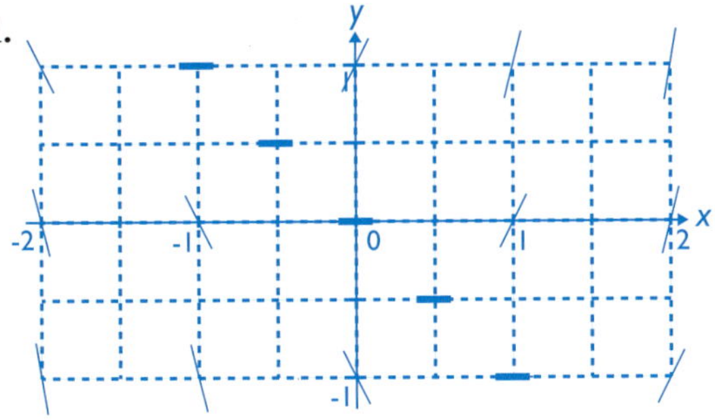

Figure SA8.1

2. **i)** d **ii)** c **iii)** a **iv)** b

3. **i)** a **ii)** c **iii)** d **iv)** b

4. **i)** c **ii)** a **iii)** b **iv)** d

5. a. This is the slope field of an implicit function.

b. The solution through (0, 0) is shown in Figure SA8.5.

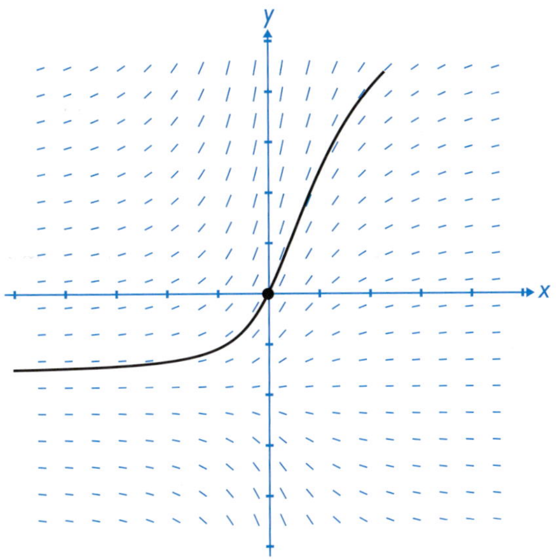

Figure SA8.5

6. a. Of y only.

b.

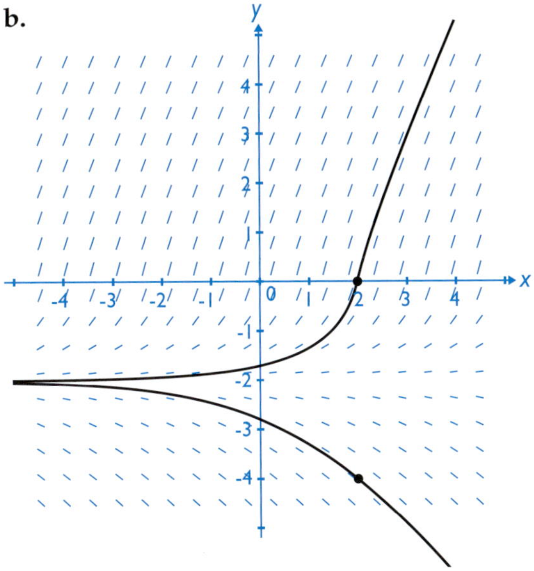

Figure SA8.6

7. Because $y^2 - 4y - 21 = (y - 7)(y + 3)$, the equilibrium solutions are $y = 7$ and $y = -3$. The graph of y' against y is a parabola with horizontal intercepts at $y = 7$ and $y = -3$ and an absolute minimum at $y = 2$. Because the parabola is positive for $y < -3$, the solutions in this range are increasing and concave down. They therefore approach the equilibrium solution $y = -3$, so that is stable.

In the range $y > 7$ the parabola is positive and increasing so the solutions are increasing and concave up. Therefore, this equilibrium is unstable.

Between the equilibria, the parabola is negative and the solutions are decreasing; they change concavity at $y = 2$, confirming that $y = -3$ is stable and $y = 7$ is unstable.

8. Substitute $y = mx + b$ into the differential equation to get $m = (mx + b) - 2x + 1$. Simplify to get $(m - 2)x + (b - m + 1) = 0$. If there is a linear solution, it must hold for all x, so that $m - 2 = 0$ or $m = 2$. Therefore, the only possible linear solutions have slope 2. If $m = 2$ then $b - 2 + 1 = 0$ so that $b = 1$. Therefore, $y = 2x + 1$ is the only linear solution to

$$\frac{dy}{dx} = y - 2x + 1.$$

9. a.

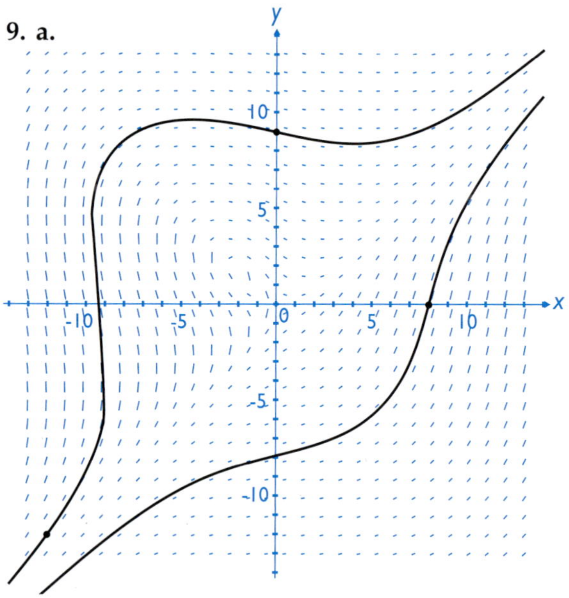

Figure SA8.9

b. The boundary appears to pass through (12, 10) and (−8, −10). The line containing these points is $y = x - 2$.

c. Regroup the derivative as $y^2\frac{dy}{dx} + \left(2x\frac{dy}{dx} + 2y\right) + \frac{dy}{dx} - x^2 = 0$, which emphasizes the derivative of a product inside the parentheses. Antidifferentiate to get $\frac{y^3}{3} + 2xy + y - \frac{x^3}{3} = C$.

10.

x	y	$\Delta y \approx y'(x)\Delta x$
1.6	2	−0.12
1.7	1.88	−0.152
1.8	1.728	−0.187
1.9	1.541	−0.226
2	1.315	